DATE DUE

| | | | |
|---|---|---|---|
| DEC 13 '82 | | | |
| | | | |
| | | | |
| | | | |
| | | | |
| | | | |
| | | | |
| | | | |
| | | | |
| | | | |

# THE LIVING OCEAN

# THE LIVING OCEAN
## Marine Microbiology

E.J. FERGUSON WOOD

ST. MARTIN'S PRESS NEW YORK

St. Martin's Press, Inc., 175 Fifth Avenue, New York, N.Y. 10010

Library of Congress Catalog Card Number: 75-15138

First published in the United States of America in 1975

Printed in Great Britain

# CONTENTS

## THE LATE E. J. FERGUSON WOOD – An appreciation.

The emergence of marine microbiology as a recognized scientific discipline was due to the foresight and industry of a small number of men, among whom the name of the late E. J. Ferguson Wood ranks high. 'Fergie', as he was known by colleagues and students alike, obtained five degrees (including a BA in English and Philosophy) from the University of Queensland. The last of these was an honorary DSc conferred on him in 1965 in recognition of his many contributions to marine microbiology.

His early work was in the area of agricultural pathology, but soon after finishing university he became interested in marine microbiology and obtained a research position at a government marine laboratory near Sydney, Australia. At that time marine microbiology could hardly be called a scientific discipline – virtually nothing was known about the subject and almost no one was working in the area. Today no marine laboratory is complete without a marine microbiological research program and research articles on the subject pour out at the rate of several hundred per year. Many of these articles contain references to Fergie's research and attest to the widespread influence he has had on the field. He himself published five books and well over one hundred scientific articles. Unlike most earlier microbiologists he saw the 'big picture' – he studied marine microbes not just for themselves, but always visualized how they fit into the grand scheme of ocean community dynamics. His work was by no means restricted to microbiology, however, for he also published on such subjects as fisheries economics, estuarine ecology, scientific English and marine pollution.

Fergie was continually generating novel and stimulating hypotheses. He had far more ideas than he and his students could possibly follow up. His writings are full of hints and suggestions for research that will stimulate other scientists and help shape their work for years to come.

R. E. Johannes
University of Georgia

1974

# INTRODUCTION BY SERIES EDITOR

The subject covered in this volume is the microbiology of the oceans, an area of study of which the author was a pioneer. The book was written shortly after Fergie Wood's retirement as Professor at the School of Marine and Atmospheric Sciences of the University of Miami. He was anxious to write a book covering the experiences of a lifetime in his field and had in fact almost completed a script, when I approached him to join the *Biology and Environment* team.

Fergie Wood's early death from cancer of the spine was a shock to his family and his many friends alike. It was also a loss to marine science by the removal of one who combined unique experience with a fresh and active mind. However, we hope that the present volume will be a fitting memorial to him and to his life's work.

The typescript we received was in early draft form and we thus faced the task of completely revising the material without altering the author's basic design or the meaning he wished to convey. This we believe we have successfully done. Following a first revision made by myself in consultation with the publishers, the script was passed to Dr. C. F. Hickling, who teased it into shape very ably. Finally, the script was passed to Dr. R. E. Johannes of the University of Georgia, nominated by the author as his Literary Executor. Dr. Johannes made some further revision and added extra material to areas we considered in need of expansion. We are also indebted to him for supplying a great many of the pictures and illustration.

In all this revision, we have had the full consent and co-operation of Fergie's widow, Mrs. Hazel Wood. I am pleased to reassure her that, in spite of all the work done on the script, this is still Fergie's book, and we hope and think much as he would have wished to make it, had he lived.

Marine microbiology might appear to the uninitiated somewhat remote from the problems of man's terrestrial environment. Just how fundamental it is to these problems will appear to those who study this book. The greater part of forming production in the sea – photosynthesis – is dependent on single-celled marine organisms, present in the upper layers of ocean water to which light penetrates. Pelagic fish and marine mammals, such as whales, could not exist without them. Furthermore, urban and industrial wastes rely on their activities to degrade them and even oil slicks – given the chance – are broken down into useful materials by their activities. An understanding of marine microbiology is, therefore, essential, if we are to understand and overcome many of man's environmental problems.

Richard N. T-W-Fiennes

# INTRODUCTION

I have enjoyed writing this book. Some of my ideas are controversial. Some of them I know to be true though I have not sufficient evidence for a conviction. My intention is that you should realise that microbes will be as important in the future of the earth as they have been in the past. They have led to our existence, and may help to prevent our destruction; they may even save us from ourselves.

As a student, I was taught that a scientist must be objective, and that every statement or observation must be documented. Also, that a scientist must seek the truth, and not be satisfied with less. It took me a long time to realize that no human being can be completely objective, and that the truth, even scientific truth, is a chimera. The most that we can get with our human minds and feeble experimentation is an approximation, and all we can hope to contribute is a better approximation than we have had to date. Newton's theories were regarded as absolute until Einstein came along; and now it seems that Einstein may be inadequate. Considering all these things, I decided that I would write a book giving the facts as I know them, speculating where necessary to connect related pieces of information. That is what I have done.

# PART I

# COLLECTING INFORMATION

# 1. THE ADVENTURE

When one is at sea in a small ship, the oceans seem endless in size and, during storms, relentless in fury. They are also very lonely. During a voyage from Kure in Japan to Dreeger on the north coast of New Guinea, some 3,000 miles, we met only one ship. It was travelling from Guam to Manila, at right angles to our route, and we had to change course to avoid a collision. I do not recall passing more than one ship between Dreeger and Sydney, Australia via the China Straits and the Coral Sea. Though the West Indies seem to form an almost continuous chain on the charts, from the Abacos in the Bahamas to Trinidad within sight of Venezuela, the distance of travel is about 3,000 miles, whether you go through the Providence Channels past Nassau and south to Barbuda, through the Old Bahama Channel past Great Inagua and Cuba or through Yucatan Channel between Cuba and Mexico. Even in this area, you might only pass a cruise ship or a freighter or two. Unless you are in the steamer lanes, and research ships tend to dodge these lanes, you are very much on your own. Research ships dodge the trade routes because they are hove to much of the time, and in rain squalls or fog, radar notwithstanding, are vulnerable to collisions.

They are expensive to run; even a small ship some 80 feet long costs about $500 to $1,000 a day; one 180 to 200 feet long costs about $2,000 to $3,000 a day, and so on. Such ships try to occupy 2 or 3 stations a day for general purposes, so every sample taken is a costly item in the research budget. Each cruise must be carefully planned, and as much information as possible must be collected from every station. Several people interested in the same type of information draw up a cruise plan. They work out the route to be followed, the 'steaming time', or time it would take the ship to cover the route steaming continuously, allowing for the speed of the ship, the weather, currents and winds to be expected and delays for refuelling, courtesy visits to other countries, and possible failures of the ship and gear.

Disappointments are frequent. I remember running from Wellington, New Zealand, to Sydney, Australia, without taking any of our 10 planned stations. The wind was on our beam, consistently at 60 knots or more. We did try once. We hove to, manned the winches and put a man in the 'chains' to attach the gear. At his first attempt, a wave took him out of the chains, poised him over the inky water, hanging onto the wire, and dumped him back into the chains. When the captain and I, who were standing by, recovered from being nearly drowned by a 'greenie', we decided to call the whole thing off. The ship resumed her course, but my assistants and I had to get the deck gear inboard to safety from the weather side. It took us half an hour, hanging on like grim death with one and sometimes both hands, and I have rarely been so terrified. Another time, in the Antarctic, my assistant was attaching a net, with the wire trailing outboard from the ship. I was trying to keep the wire straight with a 'handy-billy' when I saw a huge wave coming. I daren't let go or my friend would have gone seaward like an arrow, so we both stayed there and took it. When the wave had gone, the icy water had knocked back my parka hood, and I had a bucketful of icy water between me and my thermal clothing. We had hours of work to do before we could get below for a change of clothes and a large swig of rum.

Sometimes the gear fails. Once, on the first night out, we lost our echo-sounder and had to go into dry dock for two days. At the next station we lost our sounding-wire through corrosion though it was almost new. With it went all our best gear, and we didn't have enough wire to make the planned deep casts. Then one of our crew got beaten up at the first port, and was out of action; and we ran into impossible weather. Few people are lost at sea from research ships, though more than one ship has disappeared completely. There was the *Carnegie* with Agassiz on board, the *Endeavour* with Dannevig, and the *Fairwind*, a New Guinea research vessel, all in the Pacific. I have twice been on a ship which rolled through 90 degrees, but luckily came back right side up. We got seawater down the funnel of a 2,000 ton ship on one occasion. Getting information at sea may be difficult and possibly dangerous, however many precautions are taken.

There are difficulties, too, in getting where you want to go. Gales, storms and hurricanes may force the cancellation of a chain of stations, and this may cause a delay of a year before you can get back to that area at the right time of year for your observations to be of value. This can break the continuity of observations and may make it necessary to revise a whole programme. The weather is a very touchy factor in some confined areas such as the Coral Sea between Queensland, New Hebrides and the Solomons. Most people think of the barrier reef being a continuous reef off the coast of Queensland with clear water through the rest of the Coral Sea. Unfortunately this is not so. There are nests of reefs, some of them, like Chesterfield Reef, of large area, but discontinuous. The area is badly charted, and one may have great difficulty in getting the captain to take his ship into a confined area where sudden squalls and weather changes are likely.

Political considerations may be a factor; this happened to me when I wanted to prove an interesting theory regarding some peculiar low salinity surface waters in the western Atlantic. I had some preliminary evidence supporting the ideas of a colleague, ideas which were not only interesting in themselves, but which would have shown how well phytoplankton studies could support physical and chemical evidence of current movements. However, owing to some unfortunate contretemps for which I was in no way responsible, the Brazilian government refused to allow our research ships into their waters. As part of the area I wanted to study included the north coast of Brazil and the mouth of the Amazon, I had to abandon the research altogether, and it will probably never be completed. On another occasion, in the eastern Mediterranean, we had an Israeli on board our ship, and so could not complete a projected study of the area off the Nile delta.

Sometimes, too, ships break down. Once, in the Coral Sea, we blew a boiler and had to limp into Cairns and return to Sydney, abandoning the rest of the cruise and hoping that the other boiler would keep going. During a Caribbean cruise an engine gave out in the Old Bahama Channel and we had to try to work out deepest station, in the Brownson Deep, in rising wind and seas, abandon it half way through and limp into San Juan, Puerto Rico, to wait while new engine parts were flown in. Then we had to revise our whole cruise plan so as to get as much work done as possible in the time that was left to us. It is in cases like this that the forbearance and generosity of one's colleagues are most valuable, as everyone has to make sacrifices, and things can be difficult if one's colleagues on shore are not understanding.

Bad weather often makes it difficult to determine station positions. You may not see stars for several days and have to proceed by dead reckoning. The current information is often inaccurate, and one may find oneself in a northerly set when the chart shows a 2-knot southerly current. If you are looking for a definite position such as the Planet Trench or another of the abysses only a mile or so wide, there can be and have been difficulties in getting to the exact spot. Then, in a strong current, there is trouble in maintaining your position during a station, especially if the weather is cloudy and your station time is long – say 24 hours. One does not have this trouble if one is in an area where the captain can use Decca or Loran to fix his position with respect to the land, but even in the Caribbean there are areas where this is not possible, or where the accuracy is questionable. The most modern research ships are, of course, fitted with equipment for satellite navigation, but this is a relatively new tool and an expensive one, so all but the favoured few still have to be content with the older, less accurate means.

It should now be clear why our knowledge of the mysteries of the oceans is so meagre. Men in the bathyscaphe have been down to the deepest ocean waters at over 32,000 feet, but that was an expensive dive and no samples could be studied there or brought back. Submarines are being designed for underwater research, but they are still expensive to operate. They are being equipped to collect samples under continuous observation, and future marine microbiologists will be able to pick their samples as easily as they choose goods in a shop. Then, of course, observation will supersede speculation. In shallow waters, skin-diving and scuba-diving are being used more and more frequently, and these methods do give us a lot of information. We can actually see what is going on, watch natural processes where they occur, and collect material for study in the laboratory. Many unexpected things have been found: for instance, plants such as macroscopic algae growing in almost total darkness; but this kind of work is only in the early stages. One intrepid microbiologist has gone down through an ice hole in the Antarctic, and scraped off microscopic growth from under the ice with his flippers. One study that has been made using scuba gear is the observation of the feeding habits of mullet, which suck algae off the sea-grasses and fractionate the sediments with their fins to get the maximum organic matter from the sea bottom.

The standard methods in use today are much less spectacular. We start out with some rope or wire, rope being used only for very shallow water. To this wire we attach various devices, and lower them into the water. One such piece of equipment is the 'Nansen bottle' designed by the great explorer Nansen, who was also a great oceanographer. The Nansen bottle is still used extensively, and most other water-collecting 'bottles' are modifications of Nansen's original idea.

All this sounds very simple, but there are many hazards. First, one must realize that although things weigh a lot less in water than they do in air, weight still mounts up. Even the wire itself has considerable weight when you have a lot of it. So important is this that the Swedish research ship *Albatross* and the Danish ship *Galathea* had a tapering wire, 80 mm in diameter at the lower end and 500 mm in diameter at the upper end of the 30 miles of wire which were necessary for studying the deepest parts of the ocean. The change in diameter was necessary primarily to support the rest of the wire. Even stainless steel wire rusts and breaks when used in salt water. The tapered *Albatross* wire produced further problems in winding it on drums to keep the constant speed necessary for the working part below. There were also problems in storing 30 miles of wire when it all had to be available for use at the same time. On the *Galathea,* the winding drums were on the fantail and the wire passed through a number of sheaves to the forward magazine, where there were storage drums geared so as to take up the slack. The winding drums were tapered to allow for the changing diameter of the wire and to give constant speed.

One of the problems of the cruise leader is to calculate the weight of the gear that he will put on the wire, the strain the wire can carry, and additional strains that may be put on it by currents and possible snagging on the bottom. He usually allows for a large safety factor, but may have to shade this a bit if worsening weather makes it necessary to hurry over a critical station. As an example of current effects, the Florida Current may have a surface speed of 7 knots as it rushes out to form the Gulf Stream, but the bottom water, due to friction and other factors, will have a speed of only one knot. Also a southerly wind may push the ship ahead of the surface current. If the ship is moving north, the planner must allow a tug of over 6 knots on the wire fishing near the bottom in addition to the speed of the ship. If he is going south, the wire may be slack in the middle. Such calculations relate not only to the strain on the wire, but to the length of the wire necessary to reach a given depth. Each wire runs over a 'meter wheel' which records the amount of wire that has been run out. But the movement of the ship when hove to can allow the wire to slip on the wheel, and I have even known it to loop with disastrous results. An 'accumulator', a pulley system controlled by a strong spring, maintains a steady pull on the wire even if the ship dips suddenly. If there is any sudden slack, it should pull the wire in, even when the winch is running out.

The best winches are the old-fashioned steam ones because they generate the same power whether they are running slow or fast, and can work dead slow without any slipping. Electric winches dissipate their energy in heat when working slowly, or require a slipping clutch. The hydraulic winch is the next best, and also exerts an even pressure at all times. In small ships, especially wooden ones that 'work' in heavy seas, the pipe-lines can break, which produces an oil leak. Winchmen too can be a problem. There is always the one who doesn't wake up until the gear has reached the block and either breaks the wire or damages the meter block beyond repair. It is customary for the lookout to cry 'sight' when he sees the apparatus below the water, and 'surface' when it reaches the surface, giving a double warning to the winchman. The good winchman can bring the wire up fast and stop it every time in the right place for the 'bottle snatcher' to remove the equipment. The bottle snatcher is the man who puts on and takes off the collecting gear, and must be a very reliable person. He stands on a platform which projects outboard of the ship, usually leaning further out over a safety line of either rope or shock cord. If this breaks – and I have seen it happen – he falls into the sea.

In the old days, when our gear disappeared from view, we had to guess pretty much what was happening below. Sometimes we used Kelvin tubes to get some sort of a picture of what had been happening, but these were only of use when we got the gear back inboard. Nowadays, electronic equipment gives us a very good idea of what is going on below. Research ships have echo sounders, similar in principle to, but much more accurate than, the ones used in World War II. These are known as PDRs (precision depth recorder) and report not only the depth of the bottom, but most things that are going on in between. They record fish schools and the strange phenomenon called the 'deep scattering layer'. PDR records are made on charts or on tape, and are usually continuous so as to give a complete picture of the bottom over which the ship has cruised. They are used in the production of fair charts for hydrographic purposes.

In order to check what is happening to our 'cast' as the wire pays out, we add an instrument known as a 'pinger' to the bottom just above the weight (usually 50 kg). The pinger is a transmitter which sends out a regular ping, which is recorded audibly and graphically by the PDR. We can now see the depth of the bottom and of the pinger and the PDR is accurate to within 1 metre. In waters like those of the Caribbean, where volcanic cones and coral atolls tend to bump out of the bottom like overgrown pimples, working close to the bottom is a scary business. This applies also in the very deep trenches such as the Planet Trench in the Solomons or the Brownson Deep off Puerto Rico. These deeps are very narrow, and it is easy to touch the sides if you are not careful. You find the trench, fix your position – if the weather will allow – and steam along the trench. But there is a current, exact speed and direction unknown, a wind, possibly a trade wind, which may change in intensity during the seven or eight hours required for the deep cast, and you watch the PDR anxiously. Tempers are short. The bottom begins to shallow – do we haul up or wait and see? It shallows some more. Is this just a bump or should we move east a bit – or maybe south or west? Tension mounts, the clock moves, sweat pours. We can't stop the ship because we have more wire out than there is water below because we are towing. We have ten minutes to go and the depth is holding. The seas and wind are rising. It starts to shallow again as we give the signal to heave. We can't hurry or we shall lose the gear. We ask the captain to change course 30 degrees east into the trade wind, and this seems to work. The depth increases slightly and the wire is coming up. At last we have less wire out than there is water below. The cast is safe – we hope. The decks are wet and we are taking green water occasionally and the bottle snatcher is really getting it and the wire is jerking on the meter block. In comes sample after sample and each time we breathe a sigh of relief. Now the cast is in and we are under way again, and only the pinger remains. It has mud on it. It touched bottom. Wheew! That was a close call. Twenty-four thousand feet of water below, over thirty thousand feet of wire out and we just touched the bottom. We got the sample we wanted close to the bottom, but we were lucky.

The place where one stops to take samples is called a 'station'. Some stations are taken on the run if suitable gear is available, so a station is a point at which an observation is made, and logged. The stations are usually planned before the ship leaves port, so as to give definite information regarding certain features. Usually they give information for several scientists with allied interests. Latitude and longitude are given to the captain, and it is his responsibility to put the ship there when required.

Each time the wire goes overside, this is known as a 'cast'. The number of casts at each station is decreed by the allowable weight of gear per cast, and the number of operations involved. You usually allow 100% leeway in making a cast, e.g. if your gear weighs 500 kg in water and the wire has a breaking strain of 1,000 kg you consider it safe. Often, if you are in a hurry, or the weather is worsening, you will risk getting closer to the limit. Then, you need different casts for certain types of gear. 'Bottle' casts are made vertically, with the ship stationary, while net casts are towed, with the ship moving slowly, and electronic gear is lowered on a coaxial electrical cable, and not on the ordinary sounding wire.

## 2. THE EQUIPMENT

We now have to discuss gear, but first we must know what we want our gear to do. For our purposes as microbiologists, there are two main kinds of gear that we shall use. The earlier people, who studied phytoplankton, used fine-mesh nets which they towed behind ships at slow speeds. These nets consisted of a metal ring towed by a 'bridle' of 3 ropes which came together at a ring that was attached to the main wire. To the ring was lashed a conical net consisting of a canvas top and a silk mesh net proper, usually with 200 meshes to the inch. At the apex of the cone was a 'bucket', a glass or metal container in which the organisms were supposed to be finally caught. Similar nets, but with coarser mesh or a graded series of meshes, are still used for animals,

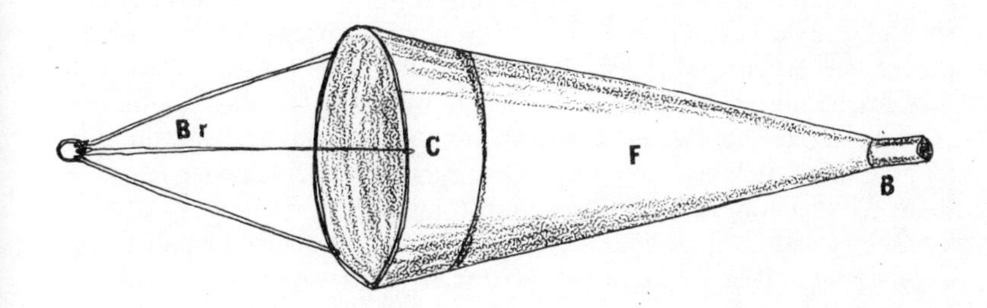


Fig. 1 Plankton net. b. bridle, c. canvas cone, f. filtering net, b. bucket.

(Fig.1) but it is generally believed that nets are useless for the smaller plants. There are two reasons for this: first, the mesh of phytoplankton nets is so fine that it is almost waterproof, in fact I have used this netting to strain water from petrol. It therefore pushes most of the water in front of it and strains only a very small proportion. The second reason is that many of the smaller microbes can pass through the meshes, fine though they are, and so are neither caught nor recognized as part of the flora. Nets come in all shapes and sizes, and each user has his own reasons for the kind of net he employs. Some nets have an inverted cone in front of the net proper, and others are cylindrical with a small cone at the lower end, and there are high-speed nets which are usually encased in some sort of metal or plastic container.

The other usual gear for catching phytoplankton is called a 'bottle' or 'trap'. Originally, glass bottles were used, and they still are in shallow waters. However, glass is a nuisance on board ship, and plastic is used wherever possible. Plastic has an advantage over metal in that metals, especially copper, are toxic to plants and animals, and this is a serious disadvantage if organisms are to be cultured. The Nansen bottle, or a modification of it, is still in use today.

At this stage, we must realize some of the peculiarities of the ocean environment. Water is almost, but not quite, incompressible, and it also has weight. A 10 metre column of water will exert a pressure of nearly 5 kg per square centimetre, or 1 atmosphere. At a depth of 10 metres there is a pressure of two atmospheres, while at 10,000 metres there is a pressure of 1,000 atmospheres. With such pressure, there are only two alternatives. You can send down a bag with all the air squeezed out, and open it at the depth you want, or you can send down an open vessel, allowing the water to flow through, and close it on the sample that you want when it gets to the right depth. Both methods are used. The usual water bottle in the oceanographic sense is a tube with both ends open and a mechanism to close them when required. Usually the tube is arranged so that the water flow is parallel to the wire, i.e. vertical, so as to minimize deflections of the wire. Closing of the bottle is done with a pair of taps or with two plugs, according to the design. Nansen used two barrel taps connected by a rod, and closed by turning the bottle over on the wire. Others use two plastic or rubber plugs, sometimes even the familiar domestic plug, closed by a stretched rubber which connects them. The vacuum type is sent down without any air inside and opened by sucking water into it by pulling the sides or ends apart like pulling on a concertina. These are preferred for studies on microbes as they can be sterilized before they are sent down, whereas the open-ended type have water going through them all the time, and some microbes may attach to the insides on the way down. The evidence is that contamination of this sort is of minor consequence for most purposes, but it does remain a possibility for critical work. The microbiologist will find that it is far more convenient to use the open bottle whenever he can as he can then share his samples with others such as chemists, by putting on a larger bottle if necessary. There are big advantages in using the same sample for as many tests as possible, as it is impossible to take two samples in exactly the same spot and so the same conditions may not prevail in 'duplicate' samples. I have found in practice that certain phenomena, including 'rafts' of phytoplankton, may be confined to a metre or two in depth, and even the rolling of the ship may make it difficult to take two samples at the same depth close to the surface. All water bottles are designed to leak, and this is necessary as water has less volume at depth so expands slightly as it is drawn up, and the excess leaks out; but there is no leak into the bottle, which has a slight positive pressure.

It may appear difficult, but the method of opening and closing by remote control is simple, and a 'messenger' is used. This is essentially a piece of metal (usually brass or bronze) with a hole in it, attached to the wire. It is slipped over the wire and allowed to fall freely. When it gets to the equipment, it hits a trigger which acts as a release and activates the gear. At the same time, it usually releases another messenger which has been attached below the apparatus, and this goes down to the next net or bottle and so on down the line. It is thus possible to put a whole string or 'flight' of nets or bottles on one wire at the same time and activate them in turn. Sometimes, two messengers are used for the same piece of equipment; the first opens or starts the mechanism and then drops off, and the second messenger closes, or stops it. The design of some of these is quite difficult, and the task of collecting gear, messengers and the like is an interesting study in itself. With a number of pieces of equipment on a wire and many messengers travelling down, there is a possibility that one or more may fail to work or work at the wrong time. When this happens, it is of course the gear nearest the bottom which does not function and this is the hardest to retrieve and takes the most time.

In planning station-times one has to allow for the time taken to run out the wire and get it back, the time required for 'fishing' a net or equilibrating other gear, e.g. thermometers, and 'messenger time' which is about 150 metres per minute. Thus a deep cast may require 45 minutes for messenger time after the gear has been run out. Nets are usually towed for twenty to sixty minutes, the longer time being required where the plankton is sparse, i.e. in the lower nets of a flight. It may take an hour or more to run the wire out, and longer to get it back, which cannot be hurried or gear may be lost. A deep cast may take 8 to 10 hours, and a deep station up to 24 hours, and there is little that can be done to speed things up. If the weather is worsening, this can be a problem, and an exercise in judgment.

We are now at the stage of electronic triggering devices, strain gauges, and some time- and pressure-controlled mechanisms. We can record the depth at which the gear is working by means of 'pressure pots', which have a stylus writing on a clock dial and activated by a Bourdon-type pressure gauge. We can use a leaking, spring-loaded piston to act as a time or pressure release by controlling the rate of leakage, or a 'strain gauge' which will break at a certain pressure or tension. Many of the wires now used for suspending the equipment are coaxial electric cables through which electronic signals can be passed, and there are even receivers and transceivers which can communicate with the deck and receive and carry out orders therefrom. An *STD recorder gives a continuous record of the temperature, salinity and depth as it is lowered through the water. The record can be marked on a sheet of paper or put on tape directly for the computer, either in the laboratory or in the head of the machine itself. The previous method of measuring temperature was by 'reversing' thermometers larger than, but similar in principle to, clinical thermometers. When the required depth is reached, a messenger allows the thermometer and its holder to turn sharply upside-down, and the mercury thread is broken. The amount of mercury then in the tube represents the temperature at that time and depth. The expansion of mercury and glass at the deck temperature is known, so the actual temperature at the depth where the measurement was taken can be calculated. Thermometers are carefully calibrated, and there are nomograms to read off the figures. In deeper waters, two thermometers are used, one 'protected', i.e. completely encased in glass, and the other 'unprotected' with one end open. The latter responds to hydrostatic pressure on the walls of the mercury bulb, the former does not, so the difference in 'temperature' indicates the pressure, i.e. the depth at which the temperature was taken. Thus, if there was a sag or bow in the wire which was not recorded by the meter wheel, the protected and unprotected thermometers would pick it up for the record. In the same case with each reversing thermometer is an ordinary thermometer which records the deck temperature, and about 15 minutes are allowed between the time that the samplers come aboard and the reading of the temperatures so that the system may come to equilibrium at deck temperature. The thermometers are read with a hand lens to the third place of decimals, as great accuracy is needed for plotting current and water-mass data.

The STD recorder works on the electrical conductivity of salt water to measure salinity, thermistors (sensitive thermocouples) to measure temperature, and pressure sensors to measure depth. The electronic output of these is transferred to an amplifying recorder, which is a specialized piece of equipment adapted from types used in industry.

An exciting development is the free-fall sampler. This consists of a sampler of the required design which is weighted to carry it to the bottom or to a predetermined depth, where it releases a weight and takes samples either going down or coming up, and then signals its whereabouts when it surfaces again. Some of these send out continuous signals and can be accurately tracked, and their information recorded on the spot. It is also possible to trigger electronically samplers attached to the STD, so that any interesting phenomenon causing sudden deflections of the STD can be studied from the retrieved samplers.

I have discussed in some detail the collection of physical and chemical data as well as phytoplankton samples because I am firmly of the opinion that biological sampling should always be accompanied by all the data that one can conveniently assemble. In this way one can carry out qualitative and quantitative studies on the organisms themselves, and give a full description of the community contained in the sample, as well as the physical and chemical conditions of the sample at that time. From this one may hope to present a picture of the relations of the plant and animal communities to their environment and their contribution to that environment. I have frequently found that my sampling programme did not include data that I really needed, and of course it is then too late to try to get those data. Many phytoplankton records from the past have to be completely ignored because of such a lack of data.

If samples from the ocean bottom are required, the two main instruments needed are the grab and the corer. The grab, as its name indicates, is a miniature mechanical grab which digs into and closes on a sample of the sediment. It takes a sample of given area, but may fail entirely if it bites on a piece of coral or a stone. In the form of the Petersen grab, it has given some outstanding quantitative surveys of the fauna of the sea-bed, even in very deep water. But microbiologists are more interested in the corer. This is essentially a tube which can be driven into the bottom by gravity (gravity corer) or by a piston (piston corer) activated by a spring, hydraulic pressure or an explosive charge. It has a valve at the top which lets out supernatant water and closes to hold the core in by vacuum. To aid this, and prevent cavitation as the core comes up, there is a 'comb' at the bottom. This is shaped like a hair comb but is bent into a circle inside the core tube so that the teeth point upwards as the core slides over them, and are pressed down by the weight of the core and close the opening when the core is lifted. The corer is lined with a long, plastic tube or 'core liner' which is removed on deck with the core attached. It can then be stored, cut up or otherwise treated. Today there are ships like the *Glomar Challenger* which can maintain an exact position in the ocean using satellite navigation (SATNAV) and transverse propellers electronically controlled by a computer, and can drill a hole in the sea bottom as well as drillers do on land or in shallow water. It can sit over some 15,000 feet of water and drill a hole 3,000 feet deep or more.

An interesting development is the use of aerial photography using infra-red cameras to study the chemistry of the oceans, at least of the surface waters. One facet of this method is the estimation of the quantity of chlorophyll in the water from aerial photographs.

So far we have been considering ocean sampling and its difficulties, but there is another, and possibly more immediately important area along the seashore and particularly in estuaries – using that word in its widest sense. In the shallowest waters, of course, you can stand and use a bucket of some sort, or a bottle, weighted and with a bung on a string that you can pull out by remote control. In deeper waters you can put the samplers on a rope or stick, or dive down with a snorkel or scuba gear and work on the bottom to take a lot of samples in a short time. To get things that attach, such as seaweeds or barnacles, you can put out glass or plastic surfaces in holders, leave them as long as you wish and bring them in at pleasure.

Aluminium beer-cans are the best markers for shallow water stations as they are easily visible, semi-permanent and unlikely to be tampered with. In deeper estuarine waters, oceanic gear can be used, but it is a good idea to make it lighter and more portable so that it can be used from small davits or by hand from small boats of about 30 feet in length, as this type of craft is most suitable for these waters.

Once the preparation, difficulties and dangers of collection of samples are completed, a study of the microbes in the water or the sediment can be made. The reason that I have gone into such detail regarding temperature, salinity and pressure is that if we are to appreciate the behaviour of the microbes, we need to know a lot about what is happening in the natural environment. So the microbiologist needs the physicist and the chemist to collect and provide him with all the data he can get. For example, certain microbes need vitamins and it was believed for some time that the so-called 'red tides' that occur on the west coast of Florida were caused by a superabundance of vitamin $B_{12}$ derived from the orange groves of the Florida peninsula. Another theory was that the phosphates escaping from the phosphate mines in the Tampa region were contributing to these red tides which cause allergies to people on the shore and kill large numbers of fish in the area.[1] Also, it is known that certain microbes are restricted to certain kinds of water, and can be used to show that this water is or is not mixed with other waters. So we have our chemical and physical data and are now ready to study the organisms themselves to try to find out what they are, what they do, and why they are where we find them. Some day we shall be able to control their occurrence, encourage the good ones and ban the bad ones.

Depending on the source of our samples, we may have too many or too few microbes to make a direct study. If there are too many, we have to dilute the sample, which is fairly easy as we have only to add a known quantity of the natural substrate (water). There are statistical difficulties here, but we are not statistically-minded at this stage. If there are too few, and this is commonly the case in the oceans, we have to concentrate the sample, and can do this either by filtration or by using a centrifuge, which may be a medical or a commercial type. Filtration has the difficulty that microbes tend to stick to the filter and some of them burst if suction or pressure is applied, and some even if they touch a filter or glass slide. In centrifugation, we make use of the fact that living organisms are slightly heavier than water, even seawater, as you can find out by trial. With an increase in gravity, produced by centrifugal force, the microbes sink to the bottom and can be collected from there.[2] Special, continuous centrifuges have been designed for plankton studies, and I have made one using most of a blender and a couple of plastic jugs.

As we are interested in microbes, we need a microscope and for the smallest ones either a very high magnification of the light microscope or an electron microscope. Ir would be too specialized to discuss here the various advantages and disadvantages of modern microscopic techniques such as phase contrast, fluorescence, interference microscopy, shadowing, or the newest tool, the scanning electron microscope. They are all extremely useful, and each has its own particular advantages and drawbacks. Some techniques are easy, some are difficult, and the ease or difficulty depends largely on the worker's experience.

There is a problem in microscopical studies of microbes, in that they do not always look the same even under the microscope. They may change form as they grow, or with changes of conditions in the environment, or for no apparant reason. It is possible to use a microscope, even at high magnifications, on board ships and in bad weather, but the observer must be wedged into position so that he moves with the ship and the 'scope, and it is a distinct advantage to be a good sailor. There are a few tricks in mounting the specimens too, so as to prevent them moving about.

A frequently-used method of studying microbes, whether marine, aquatic or terrestrial, is to culture them on artificial media which may or may not be based on the presumed natural environment. During the early stages of microbial, particularly bacterial studies, it was considered the correct thing to grow each organism in a 'pure culture', i.e. to isolate it from other organisms. In medical studies such as those of the pioneers, Pasteur, Ehrlich, Koch and others, this was an advantage, as pathogenic microbes could be studied on blood or bile media according as they were suspected of causing diseases of the blood or the intestinal tract, of humans or animals. However, the medical techniques were adopted uncritically for studies of microbes which were not pathogenic, and which lived under natural conditions with a number of other organisms, microbial and others. Beijerinck and Winogradsky, two students of soil bacteria, rebelled against this dictum, and tried to study soil microbes as part of a soil system, or ecosystem as it would be called nowadays. Their studies were very successful, and they came to some important conclusions which are still regarded as valid. I use their techniques for demonstrating to students just how each microbe prepares the way for other microbes, and their interdependence, also that there is a cycle of events which are caused by or cause the predominance of certain organisms. However, there are still people who consider that pure cultures are the only method of study.

The best method that I have found for studying marine or freshwater microbes is to look at them in their natural environment under as many different conditions as possible, then to try to isolate them in reasonably pure culture, test their limitations, and finally to put them together one by one under controlled conditions so as to duplicate as far as possible in the laboratory the original conditions. This routine is very similar to what are known as Koch's postulates for pathogenic bacteria. Koch, who discovered the diphtheria bacillus, decreed that to establish the bacterial cause of a disease he had to isolate the organism in pure culture from diseased tissue, and show that it produced the disease, usually in laboratory animals.

These processes sound easy, but are in fact difficult in most cases. Many of the microbes will not grow under their 'natural' conditions in culture, and some will not grow at all out of their natural environment. I thought at first that this was due to lack of technique and ability, but revised my opinion when I found my colleagues in the same difficulties. There have been limited successes, but mostly in greatly enriched media and under different conditions, so that even the most successful culture experts cannot say what the organism does in its natural environment. On the whole, culture studies, though such an obvious method, have not been startlingly successful in explaining the natural growth of marine organisms, and no one can tell why.

Another method, not too successful so far, is to look for changes in the natural environment which can be attributed to certain microbes. Usually such results are equivocal. For example, it is believed that a blue-green alga, *Trichodesmium* (Fig. 5), can fix nitrogen from seawater. However, it has not been possible to separate it from other microbes such as bacteria which may be the real nitrogen fixers. Moreover, if this organism does fix nitrogen, it is the only marine member of its group known to do so, and it does not contain certain structures which have been otherwise associated with nitrogen fixation among the blue-green algae. So there is still doubt about this important matter.

An important help to the microbiologist today is the computer and, although there are still limits to their usage, computation and tabulation which previously would take days or months to work out can now be carried out in seconds.

*Salinity, temperature, depth.

## NOTES


1. Gunter, G., Williams, R. H., Davis, C. C., and Smith, F. G. Walton. 1948. Catastrophic mass mortality of marine animals and coincident plankton bloom on the west coast of Florida, November 1946 to August, 1947, Ecol. Monogra. *18*: 309-324.
2. Wimpenny R.S. 1966. *The Plankton of the Sea.* Chapter XI: Faber & Faber, London, 1966.

# PART II

# THE MICROBES OF THE SEA

# 3. PERSPECTIVE AND CLASSIFICATION

But for microbes, you and I would not exist: if they were not still around in sufficient numbers, we would not remain here long. They play a large part in predigesting and digesting our foods; we all know about our intestinal flora of bacteria, and most of us have an idea of fermentations such as those connected with cheese manufacture, yoghurt, and the production of alcohol. In Japan, and other eastern countries, marine bacteria and fungi play a large part in the partial fermentation of seaweed and fish products. Microbes are important to us wherever they occur.

We find it difficult, and often impossible, to grow many important microbes in test-tube cultures, and still more difficult to reproduce in their entirety conditions in the oceans. Many people ignore this, and even well-known marine microbiologists have been known to produce some odd statements in consequence.

I am interested in the microbes of the sea, not only because of their importance now, but also because they are trying to tell me things about the dawn of life, when many of them existed much as they do today. They show me beyond peradventure that conditions of life have altered very slowly and continuously through the geological ages, and that ice ages, huge volcanic eruptions, major twisting of the earth's crust (orogenic and tectonic movements) were minor events compared with the progress of life.

The literal definition of 'microbe' derived from the Greek, is 'a small piece of life'. It is perhaps too vague and needs qualifying, but my definition will still be a broad one – 'any organism which requires a microscope to observe it'. It usually consists of one or a few cells, may or may not be able to move of its own accord, and may be able to live either as a plant or an animal, or frequently as either, as the situation demands. A microbe can go to the moon and back in a spaceship without other assistance, and may not even require air to breathe. In shape and size it may represent the ultimate in simplicity, but its way of life may be exceedingly complex. One substance that all microbes require is water, and that is why it would be impossible for microbes to exist on the moon. Some microbes can exist for a time when desiccated, but even then only at limited temperatures, below 120° centigrade. The earth is an aqueous system, and the chemistry of the earth is a water chemistry.

Living matter on earth is dependent on the ability of carbon atoms to join each other in short or long chains, spirals, rings and complex meshed structures, at the same time adding on hydrogen, oxygen, nitrogen, sulphur and combinations of these, together with an occasional atom of a metal such as iron or magnesium. This has given us the discipline of organic chemistry, with a vast range of organic compounds, the characters of which fill many volumes. The chemistry of *living* plant and animal matter forms the subject of what is now regarded as a distinct discipline, biochemistry.

It is interesting to know that the biochemistry of living plants and animals is dependent on the inorganic chemistry of our surroundings, that is, on the chemistry of non-living matter, and on the equilibria of inorganic chemical equations. When a footballer runs out of energy he is suffering in part from an unfavorable potassium balance. Heat exhaustion is combated by taking salt (sodium chloride). Even rarer trace elements such as copper and cobalt may restrict life. The Coorong, an area of southern Australia, was known as the Ninety-mile Desert until it was found that the land became extremely fertile merely by the addition of traces of these two metals.

The salts of the blood are essentially in the same proportions, but not concentrations, as in seawater, and both blood and seawater are well buffered to keep the pH constant. The *pH* is an expression of the alkalinity or acidity of a solution, and normally ranges from 1 to 14, with neutrality at 7; below this representing acid, above it alkaline conditions. *Buffering* is the capacity of certain substances to keep the pH constant at an appropriate level by giving acidity when the solution becomes too alkaline, and alkalinity when it becomes too acid. Phosphates and carbonates are excellent buffers, hence the use of these in proprietary medicines to combat acid stomachs. A microbe cannot be said to have an acid stomach, but it needs to live in a buffered solution.

Inorganic substances such as nitrogen in its various forms, carbonates, phosphates, iron salts, potassium, sodium, calcium, magnesium, and a host of lesser elements play a major part as nutrients of our microbes, and some of these also, by changes in concentration and by variation in different parts of the oceans, govern the distribution of many microbes and their movements within the ocean.

The importance of our subject – microbes of the sea – lies in the fact that the large part of life in the sea consists of such microbes, or, as they are more technically called, micro-organisms. Whereas the land possesses forests and grasslands, the sea, apart from the massive kelps around the fringes of land in colder waters, the sea-grasses of many estuaries, and the smaller seaweeds of warmer waters and coral reefs, has a plant life consisting almost entirely of tiny plant cells from 1/10,000 inch upwards in length. Many of them are not living strictly as plants, but are biologically analogous with the grasslands. They are the cells on which sea animals feed, and which must, unless we have perpetual motion, outgrow all the marine animals put together.

All animals on land or sea are dependent either directly (herbivores or 'plant-eaters') or indirectly (carnivores or 'flesh-eaters') on plant material for food. The herbivores, like cattle on land or mullets in the sea, feed directly on the plants, while the carnivores, such as lions or tuna fish, feed on the herbivores. We have here to consider what might be called the 'garbage pail effect'. At each feeding level (called 'trophic level') there is a wastage, resulting in the production of faeces, loss of energy through futile motion, so that more plant material must be consumed than would be calculated as the minimum requirement for the carnivore. When we have a longer food-chain, i.e. two or three carnivores feeding on each other in turn, the loss can be enormous. A characteristic of the sea is that food-chains are longer than on the land. On land, the carnivore – lion, tiger, etc. – feeds directly on the herbivore – cattle, wildebeeste, etc. – but in the sea, it is usual for various shrimp-like animals such as *copepods* to be herbivorous, while others may be carnivorous, e.g. the *euphausids,* the larger crustacea and many fishes feeding on the carnivorous shrimps or smaller fishes. Thus pilchards or sardines feed on microbes, and tuna feed on pilchards or on the carnivorous euphausids which feed on copepods. The creatures in each link of the food-chain have to find the energy for their own lives, and cannot increase in weight and quantity, until this maintenance requirement is satisfied. This is believed to amount to about 90 per cent of the energy they take in. So only about 10 per cent goes to growth, which in turn provides the energy for the next link in the chain. Recent experiments suggest that the efficiency may be increased somewhat by a number of side reactions in what has come to be called the food web, since there are some feedback mechanisms.

Feedbacks occur all through the food web, and many occur entirely among the microbes themselves, where the tiny, animal-like protozoa devour the plants, and where the bacteria, also to be numbered among the microbes, feed on the dead and the dying, both large and small. Some of the largest animals in the world, giant whales and manta rays, feed on many of the small planktonic crustacea.

There are a few organisms which cannot be classified as either plants or animals, and now things are more complicated than ever in this area; plants, or some of them, can live like animals, and both plants and animals can live like bacteria under certain circumstances. Some bacteria even have a form of chlorophyll which they use in a peculiar way. Some of the plants which have a dual role may be cannibalistic, and one finds cells of Dinoflagellates ( a group of phytoplankton organisms) with other Dinoflagellates inside, being digested. However, the fact that a plant cell is inside an animal does not always mean that it is a victim; it may be a welcome lodger. If it were not for the microbes which destroy and transform dead animals and plants, the sea would by now be full of the dead bodies of fish and other organisms.

The problem of separating plants from animals lies not with them but with us. Human beings pigeonhole each other and the things in their environment. Even in mathematics we cannot envisage a continuous series; it must proceed by a series of jumps: n, n+1, n+2 . . . and so to infinity. Infinitesimal calculus, as its name implies, tries to make these jumps infinitesimal, but they are jumps nevertheless. It is the only way we can comprehend infinity. Nature, on the other hand, has no such limitations, and we get phenomena which appear to be continuous. We decided in times past that chemical processes were immutable, that there were and could be only 93 elements, and it was a shock to the chemists when they found a new system of elements which they called the 'rare earth elements'. Now we are accustomed to radioactivity and the transmutation of metals, radium to lead, carbon 14 as well as carbon 12, and so on, but we are still distrustful about it. With regard to plants and animals, we are still less happy with the idea that they do not fit into a preordained pattern. Linnaeus originated the idea that plants and animals could all be divided according to their lineage into classes, orders, families, etc., and proceeded to outline what has become known as the binomial classification. Each animal and plant is assigned to a definite genus and species, is given a generic and a specific name, and these two names serve to identify the particular organism to anyone who meets it. It is also believed to indicate the line of descent of the organism, or 'phylogeny' as it is termed. Thus, when we give a generic and specific name to an animal or plant, even a single specimen, we believe we are describing what it is and its parentage right back through the ages of evolution. The name of the author who described the species is attached to the genus and species, e.g. *Licmophora* (genus) *abbreviata* (species) Agardh (author), the genus and species being usually written in italics and the author's name in Roman.

Frequently two apparently very similar life-forms have evolved in different ways, and have a different history, while others which appear superficially different are in fact closely related. In the case of the microbes, one form will change to the extent that its offspring may seem to belong to different genera, while another, apparently closely related, will remain the same under all conditions. One organism which I encountered even had two important parts which placed it in different families according to the way you looked at it.[1] On one occasion, I had a very pretty culture of an organism which I called *Gymnodinium mirabile*, of which I was studying the nutrition as well as its use as food for certain animals. My assistant had made a number of cultures from single cells, and I was looking at them one day under the microscope when I detected an apparent intruder of another genus, a *Gyrodinium*.

Both forms appeared in every culture, and further experiments showed similar changes; in fact, there were some five forms which had been described by others as different species but derived from the same pond. It had variable feeding habits too so I had to abandon its use as a test organism. On the other hand, an organism which I had isolated from the same place, and which I called *Gymnodinium simplex,* was always constant in form and nutrition, and behaved perfectly as a test organism.

These are simple instances of the life history and behaviour of marine micro-organisms. Many microbes that grow abundantly in the sea have failed to grow in the laboratory, even under the most favourable conditions.[2] One of these is a blue-green alga called *Trichodesmium* or *Oscillatoria* (there is some argument as to the correct name, and to the number of species of this genus that exist). Except when very sparse, this 'blue-green' alga is coloured brick-red to orange, and this is not uncommon among this group. *Trichodesmium* is abundant in the tropics and is believed to fix nitrogen dissolved in the water, much in the same way that root-nodule bacteria fix nitrogen in the roots of leguminous plants, such as clover on land. One student claimed to have grown this species in culture, but the material he sent me contained other organisms which showed that he had scooped up the material from the sea, and had not cultured it at all, and indeed, no one has succeeded in growing *Trichodesmium.* However, Ramamurthly[3] cultivated it to the point where he showed that *Trichodesmium erythraeum* did not respond to inorganic phosphate or nitrate but gave an 8-12-fold increase in growth in response to 2 mg/l of gibbellic acid. Even when microbes do grow in culture, it is very difficult to reproduce truly natural conditions, and, like human beings, microbes behave differently in a crowd than in their natural environment. One even finds that some microbes are eaten by animals when they are relatively sparse, and not only rejected but actively avoided when they are numerous. Dr. Alister Hardy of Oxford has his Animal Exclusion Theory, which documents the avoidance of phytoplankton blooms by animals[4]. *Phytoplankton* is the name given to the microscopic plants which float and swim about in the water, to distinguish them from the other microscopic plants which may also move about but which are normally attached to larger plants and bits of debris, or even occasionally to such things as whales and are called 'epiphytes'.

The study of marine micro-organisms goes back to O. F. Müller, about 1780, and until the latter part of the nineteenth century was mainly taxonomic. Ehrenberg, round about the 1830s and 1840s, did conjecture about the significance of microbes in the geology and biology of the earth, but few dared to follow him in his speculations. About this time, men began to use microscopes, and study the tiny organisms which were to be found in chalk-pits, on beaches, in kieselguhr (diatomaceous earth), and other interesting materials.

Because of the difficulties of classification, a large number of species was created, some of which have no real existence. This has made it very difficult to identify the kind of organism which lives in a particular environment, and why it lives there. Also, you may want to know if it is related, and how closely, to another organism living in the same or another location. The tendency in taxonomy is to choose a certain outstanding characteristic and to use it as a base, assuming that it does not vary or that variations can be discarded. Any change, then, represents another species or genus. As already noted, some organisms conform, while others do not. I have a theory that certain microbes occurring in the Antarctic close to the continent or in the ice are identical with some which are to be found in estuaries in warmer waters, and also occur in boreal (northern polar) regions. There is other evidence to support my idea that the Antarctic continental waters behave like those of an estuary, and that present distribution of the microbes is related to the ice ages. One taxonomist is very anxious to put all Antarctic species in separate groups, even separate genera; if he is right, this theory falls apart. Hence the importance of correct taxonomy in the study of the distribution of organisms.

Further difficulties arise because some characteristics that appeared identical under the light microscope seem very different under the electron microscope, and there is a tendency to use such characteristics to create still further species. The more recent scanning electron microscope shows a still different series of pictures, some of which cannot yet be correlated with previous observations. If one could make cultures at will, or observe changes in the field by a series of well-regulated experiments, one might be able to resolve some of these problems. In the meantime, one is very often left guessing. This is important, because, for example, there are some pairs of species which are morphologically almost, if not quite, identical, one of which is toxic, the other not. So, when you find a presumably toxic species growing happily with the things it is supposed to poison, are you to assume that it is a non-toxic strain or a different species? The difference is important also because if it is a non-toxic strain one might use it to crowd out a toxic one from a given area, and thus establish biological control. If it is another species, it is of no value.

In the latter part of the nineteenth century, a number of expeditions collected microbes, including plants and protozoa, from the several oceans. Names of such expeditions that come to mind are the Novarra, Challenger, Valdivia, Scotia and Siboga, and in this century, the Dana, Carnegie, Discovery, Albatross, Galathea, and the numerous cruises by ships from America and European research stations. Others collected from merchant ships and on various voyages as they toured the globe. There resulted a series of studies of geographical distributions, and it was found that some species were very limited in the area in which they occurred, while others were spread throughout most of the oceans and their offshoots. At this time there came into being the concept of 'indicator species', i.e. species occurring in limited areas, so that their unusual occurrence outside this area indicated that there was a connection, perhaps ephemeral, between the two areas.[5] At least, it gave rise to some interesting speculation, and is still used in a modified form with computers to delimit distribution, degree of association of species, and significance of their appearance or non-appearance. As the geographical studies developed in the present century, and oceanographic expeditions became more numerous, our knowledge and interest grew, and it became possible to relate certain species to certain types of water. Here began an ecological approach which is still continuing with more advanced ideas and methods.

Apart from pure taxonomy, the microbiology of onshore environments, including estuaries, has been mainly inspired by economic considerations such as the microbial control of sewage pollution, fouling of ships' bottoms, corrosion, and, latterly, pollution in various forms; and phytoplankton studies in regard to fish culture. My own chief interest in marine microbes is in their ecology. Here we first find out what living organisms occupy the marine ecosystem (taxonomy and zoo – and phytogeography), then what their relationship is to the environment, and finally why they are there. This requires study of the physiology of the organisms in the system, their biochemistry, and the characteristics of their surroundings, both physical and chemical.

Some bacteria were shown by Pasteur to cause disease, and evidence of interest in marine bacteria was shown in the 1880s when cultures of seawater were made by Fischer and others. The greatest interest in bacteria associated with marine environments lay in the effect of seawater on pathogenic bacteria, particularly those causing diseases in humans. Most coliform bacteria – the bacteria of the human intestine – are normally present in our systems, and are not pathogenic, but they may be accompanied by harmful bacteria in lesser numbers, such as typhoid, cholera or dysentery bacteria. The coliform bacteria are therefore used as an indicator of the dangers of human pollution, but the numerical relationship between these bacteria and the pathogenic ones depends on the incidence of intestinal disease in the local human population. Thus in a region where the occurrence of typhoid, dysentery or hepatitis is low, a high coliform count might be ignored; whereas if these were high, a low coliform count would be very significant. It has been found that seawater tends to kill off coliform bacteria quite rapidly.

The phytoplankton became an important study in the latter part of the last century when fishery biologists realized the significance of these microscopic organisms in the economy of fisheries. The significance of the bottom microbes (benthic) and attached organisms (epiphytes) is still not fully realized, especially in the estuarine environment.

The most primitive organisms are, of course, the viruses. In fact, there was a long argument as to whether they were to be regarded as organic chemical compounds or as living organisms. They are very simple in structure, but can reproduce. They seem to be able to survive on their own for a time, but require a host to grow and reproduce, and must be classed as parasites. They do not seem to be abundant or common in the oceans, but have been recorded from time to time. The bacteria (Fig.2) are next in order of complexity, and while containing

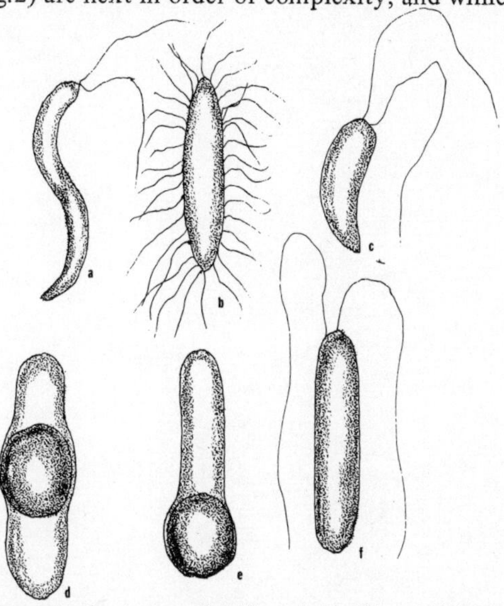


Fig. 2. Various marine bacteria. a. Spirillum, b. Bacillus, c. Vibrio, d. sporting bacillus with central spores, e. sporing bacillus with terminal spores, f. Pseudomonas. a, c and f have terminal flagella. b. has peritrichous flagella.

a number of chemical entities within a membrane, do not have a visible nucleus. A few of them contain, as we have said, a primitive kind of chlorophyll and have an interesting type of photosynthesis. Bacteria are classified mainly on their shape and their ability to use various substances for their growth, and this, in the hands of uncritical workers, has led to great confusion. The American freshwater bacteriologist Henrici has said that indigenous marine bacteria are transformers of materials, and this is a fair statement of their role.

The plants, acting as such, are active in the construction of organic matter, using photosynthesis to produce it from inorganic materials, or to make more complex material from simple organic matter. The animals act as burners, using oxygen and giving off carbon dioxide just like a fire does, though not as rapidly.

The plants may be divided into three main sections: the diatoms, which were regarded as the most important plants in the seas; the flagellates, which were regarded as secondary (often erroneously); and the blue-green algae, which were thought unimportant.

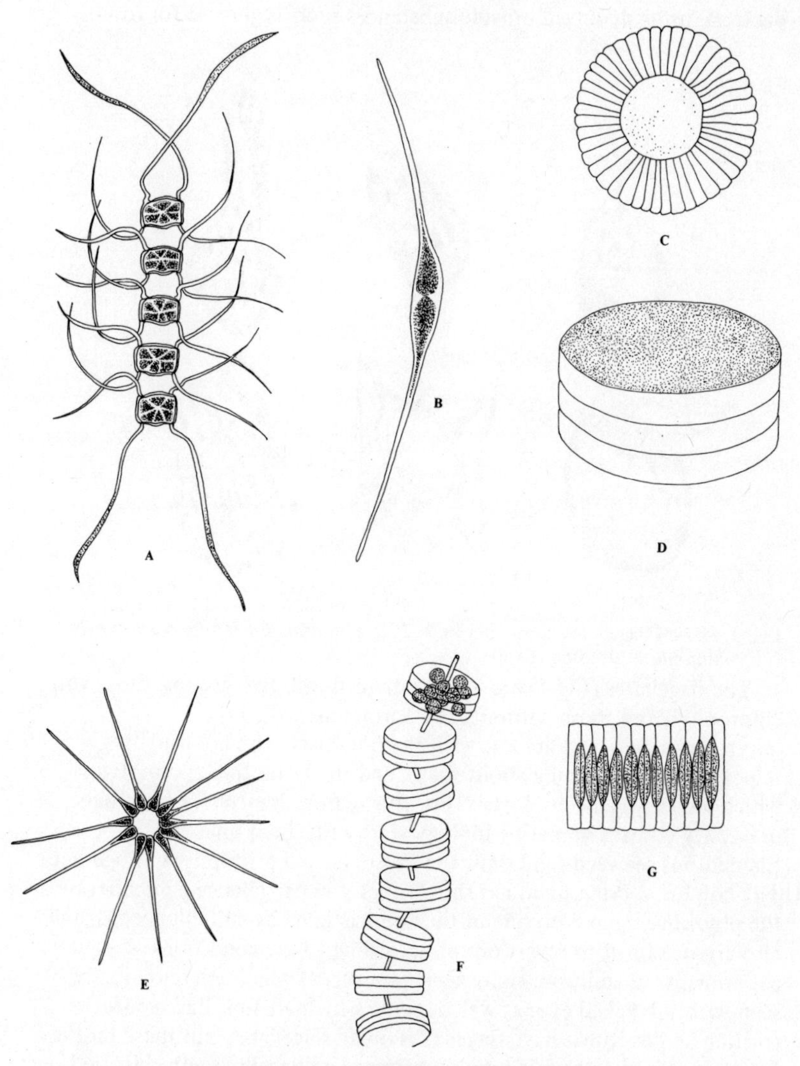


Fig. 3. Various marine diatoms. A. *Chaetoceros*, B. *Nitzschia*, C. *Planktoneilla*, D. *Coscinodiscus*, E. *Asterionella*, F. *Thalassiosira*, G. *Fragilaria*

The diatoms (Fig.3) are characterised by having an external shell (also called a 'test' or 'frustule'), made of silica, and usually very beautiful, with tiny holes or chambers. Some are mobile, and seem to have several ways of getting about. They are essentially plants, but some can live like bacteria, using dissolved organic substances such as glucose for food.

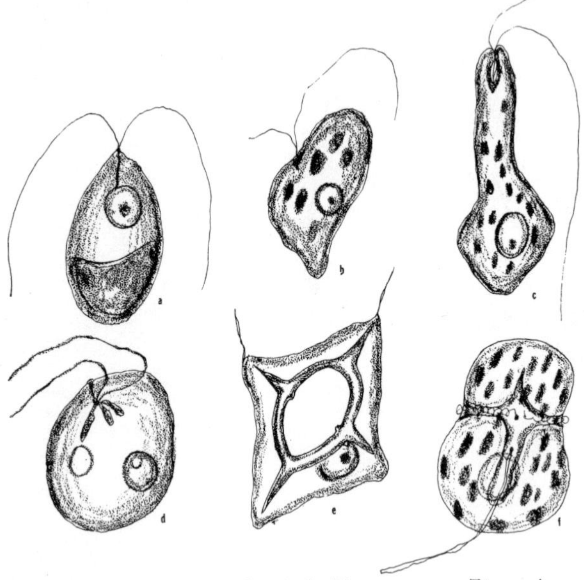

Fig. 4. Marine flagellates. a. *Isochrysis*, b. *Platymonas*, c. *Eutreptia*, e. *Dictyocha*, f. *Gymnodinium* (dinoflagellate).

The flagellates (Fig.4) are usually divided into two groups, those with chlorophyll and those without. The former are called the 'phytomastigina', and are left with the botanists, who are usually reluctant to do anything about them, and the latter the 'zoomastigina', which are left to the zoologists. For this reason, we usually find that university courses in marine biology start with the sponges or the filamentous seaweeds and omit the protozoa and protophyta altogether. The late Ernst Pringsheim has shown that you can change a member of the phytomastigina into one of the zoomastigina by suitable means, and also arrange for it to revert or not according to the conditions of your experiment. In addition, some algae (seaweeds) which are quite easily seen with the naked eye, as well as some which are not, have spores or fruiting bodies which have flagella, as do all flagellates, but these motile forms are merely stages in another group. The flagellates are classified essentially on the number and arrangement of the flagella.

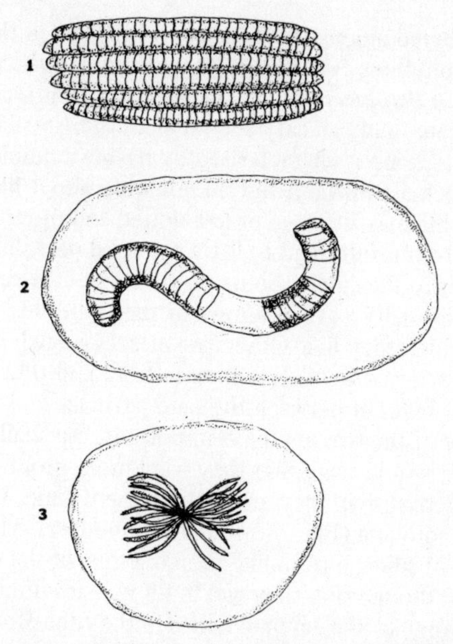


Fig. 5. Blue-green algae of the phytoplankton. 1. *Trichodesmium,* 2. *Katagnymene,* 3. *Haliarachne.*

The blue-green algae (Fig.5) are the most primitive of the marine plants, and there have been some suggestions that they be included with the bacteria, with which, in some ways, they appear to intergrade. They have no defined nucleus, and the chlorophyll is spread throughout the cell, and not enclosed in a membrane to form a chloroplast as it is in all other plants. They are the 'pensioners' of the marine world, and can live in marginal conditions, probably because they are the oldest known plants.[6]

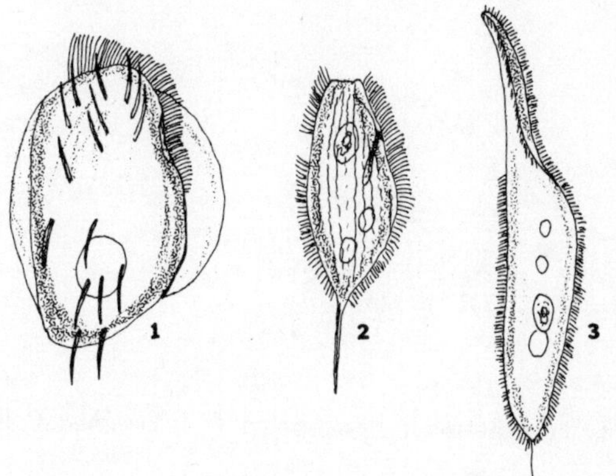


Fig. 6. Ciliates. 1. *Euplotes,* 2. *Euronema,* 3. *Lionotus.*

Also to be numbered among the protozoa which live in the sea and do not have chlorophyll are two groups, the ciliates and the amoebae. The ciliates (Fig.6), or *Paramecium,* may be very important, can live under extreme conditions, and can eat bacteria and other small organisms voraciously. This group is characterized by having a number of 'cilia'-like bristles, which are movable but do not whip about like flagella. Some of these cilia may be more or less united and move in concert, and all are used to give motility and to drive the food past the 'gullet' where it can be taken into the ciliate. Some of the ciliates are to be found in the plankton, especially a group known as the tintinnids, which have a sort of cup in which they live. Others are attached, such as *Vorticella,* which is often used in school class material; and still others live in the sediment and dash about between the sand particles looking for food.

The amoebae of the seas are quite important, especially to the geologists, and it would seem that they were more prolific and important in the past than they are at the present time. The two major groups are the radiolaria (Fig. 7) and the foraminifera. The former have skeletons made of silica, but, unlike the skeletons of the diatoms, these are internal, and though tiny they can build vast accumulations of sediments like some of the Jenolan Series in the Blue Mountains of New South Wales.

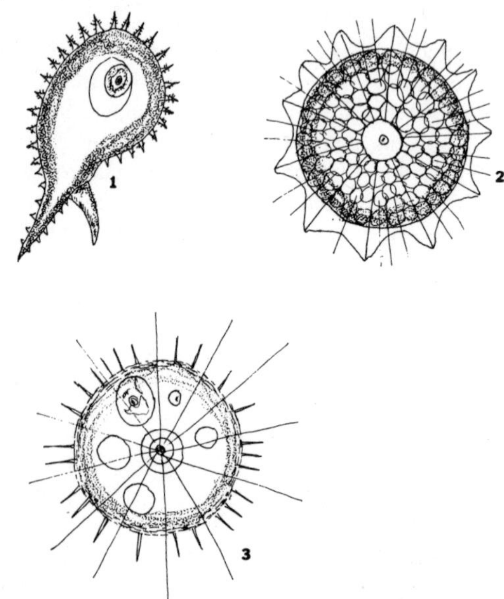


Fig. 7. Planktonic radiolaria. 1. *Challengeron,* 2. *Aulacanthus,* 3. *Acanthocystis.*

The foraminifera (Fig. 8) have skeletons of lime (calcareous) and are probably more important at the present time than the radiolaria. They usually form little whorls of cells attached to one another, and frequently look not unlike tiny shellfish. If you look at an Admiralty or U.S. Hydrographic chart, you will see the depths given, and the mysterious words *glob. oz.* This is to be translated as *Globigerina ooze* (*Globigerina* is a genus of foraminifera), and means that the bottom is covered with skeletons of these organisms.

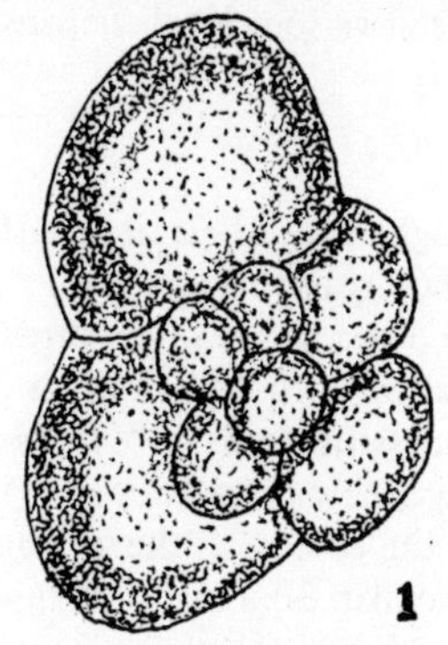


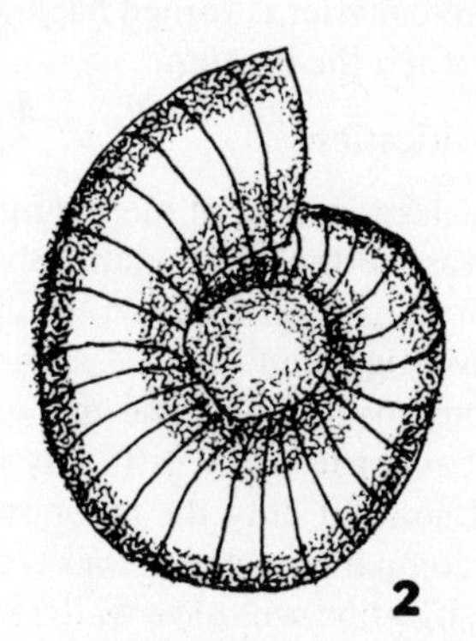


Fig. 8. Foraminifera. 1. *Globigerina*, 2. *Operculina*.

We have now considered some aspects of microbial activity in a general way. I have found some of the microscopic plants complete with chlorophyll in the depths of the oceans where sunlight never penetrates.[7] Chlorophyll is the substance that enables plants to use the energy of sunlight to produce living matter. In fact, if it were not for the activity of chlorophyll and the plants that contain it, life on earth would not continue for long. Why these plants continue to live away down in the deep oceans is not known, and how they manage to sustain themselves is a mystery. We do not know in any detail the interaction between the microscopic plants and the microscopic animals nor between these two groups and the bacteria and fungi. Studies on fungi in the open sea have been sadly neglected, and our limited knowledge has created more problems than it has solved.

About 40% of the ocean floor consists of a red clay, often high in manganese which forms large nodules. These are believed to be of microbial origin, though I cannot see how this could be so. As well as the *Globigerina ooze,* there are deposits of shells of diatoms (diatomaceous ooze), and radiolarian oozes which solidify to form beds of chert or jasper. In some places there is a fairly high organic content, especially in the deep trenches and close to the continental shelves, and in such areas there is usually an active bacterial population which serves as food for many animals that live down there. The amazing thing is that large quantities of organic matter (detritus) do not accumulate on the bottom of the sea, except in shallow waters. This can only be because the microbes in the water attack and break down such material before it can get to the bottom. It has been reported that there are often large concentrations of bacteria on isopycnal surfaces, i.e. the planes between water masses, and, no doubt, these act as border guards, preventing much of the organic matter from escaping. So nearly all the organic matter is turned back into the water at various levels before it can reach the bottom.

**Classification**

The classification of marine microbes is fraught with difficulty, partly because of the close relationship between the organisms and their environment, in that each can influence the other more or less directly. I have suggested that the Linnaean classification with its binomials is not by any means ideal, and that certain other classifications have been and are being tried. The difficulty in introducing a new classification is that it affects not only the taxonomist, but also the ecologist, the geologist, the comparative physiologist, and the biochemist. So a major change in classification will have wide repercussions, and may affect the interpretation of data in several fields of scientific endeavour. The skeletons of diatoms, for example, are common in fossil beds, and are used by geologists to trace the history of such geological strata. If characters other than those of the skeletons were adopted as a basis of classification by the microbiologists, it would cause confusion to the geologists and produce a schism between them and their biological colleagues. We therefore have to accept taxonomic principles which we regard as inadequate, unless we can provide a new system which raises no objections. With the dinoflagellates, the geologists have developed their own classification for the fossil forms, in order to avoid arguments with the biologists. This has disadvantages if we want to equate the fossil environment with that of present-day dinoflagellates.

The present-day classification of microbes was initiated and has been maintained using the light microscope; in fact one must wonder that the older microscopists could discern so much detail with the primitive instruments that were available to them. Recently, one group, the coccolithophores, has been re-studied using the electron microscope, and some new divisons have been made. A new tool, the scanning electron microscope has now become available, which shows features that are not obvious with the light microscope or the electron microscope; as it too is used in developing or revising the classification of such microbes as the diatoms and the coccolithophores, it will be very difficult to relate the new structures to those visible by present methods. Perhaps this would be desirable, because a complete revision of all the microbial groups and a simplification of present taxonomy might result.

The accepted division of the plants and animals of the world is into kingdoms, sub-kingdoms, classes, orders, families, genera and species (or their equivalents). We have seen that there are difficulties in assigning certain microbes to either the plant or animal kingdoms because they have the characteristics of both, either at the same time or alternately. We have found also that there exist in nature cells which have the characteristics of two families, and we have cultured microbes that produce individuals taxonomically ascribed to two genera. It has been generally accepted that cross-breeding is possible between two species of a genus, and more rarely between genera. For example, the zebra looks like a striped horse, but these two genera cannot produce viable offspring. Nor can the horse and the donkey. We know that the genetic and chromosome make-up of the nucleate microbes is essentially the same as that of the higher plants and animals, and can assume that this also holds for the microbes without organized nuclei – the bacteria and the blue-green algae. So I think we must conclude that the taxonomic divisions in the microbial world do not have the same significance or validity as they do in other biological groups. So I shall use groupings for convenience, but without laying any stress on their taxonomic validity.

The groups of organisms that we must consider include the bacteria, the fungi, the diatoms, the flagellates, the ciliates and the amoebae. We shall have to divide the flagellates into coloured and colourless, and the coloured ones still further into dinoflagellates, coccolithophores, chrysomonads, cryptomonads, xanthomonads, chlorophyceae and the euglenids – a large group of organisms.

This said, we may proceed to study the groups important to our subject.

## NOTES


1. Ferguson Wood, E. J. 1962. 'An unusual diatom from the Antarctic.' *Nature 184:* 1962-3.
2. Ferguson Wood, E. J. 1967. *Microbiology of Oceans and Estuaries.* Elsevier Publishing Co, Amsterdam.
3. Ramamurthly, V.D. 1970. 'Experimental study relating to red tide.' Mar. Biol. *5* (3): 203-4.
4. Hardy, A.C., and Gunther, R.E. 1935. 'The plankton of the South Georgia whaling grounds and adjacent waters.' *Discovery Reports.* 11: 1-146.
5. Russell, F.S. 1939. 'Hydrographical and biological conditions in the North Sea as indicated by plankton organisms.' *J. Cons. Internat. Exp. Mer. 14* (2): 171-92.
6. Stewart, W. D. P. 1966. *Nitrogen Fixation in Plants.* University of London, The

Athlone Press, 1966.
7. Ferguson Wood, E. J. 'Diatoms in the Ocean Deeps.' *Pacif. Sci. 10* (4): 377-81.

# 4. BACTERIA, VIRUSES AND FUNGI

The old division of microbes into 'producers' and 'consumers' is no longer tenable. The producers in the sea were the diatoms, small plants which were called 'the grass of the sea', and a few other groups which were far less important. The consumers were the bacteria, and were afforded much less study.

Dr. Claude ZoBell, the pioneer of marine microbiology, was concerned primarily with the bacteria and their role in the oceans, especially in the deeper waters.[1] He began to study the possibility of bacteria living under pressure, and to classify the bacteria which he found there and in the shallows. Bacteria have few characteristics of use for identification. They may be round (coccoid), rod-shaped (bacilli), bent (vibros) or twisted (spirillá); they may have organs of motility (flagella) and stain red or blue with the differentiating Gram stain, which reveals the structure of the bacterial cell-wall. Cocci, such as *Staphylococcus* and *Streptococcus,* are Gram-positive and stain blue, while intestinal bacteria like the colon bacteria (coliforms), typhoid, dysentery, etc., were found to be Gram-negative. The comma-shaped bacteria known as *Vibrio* (including the cholera bacterium) and the spiral forms (*Spirillum*) are also Gram-negative. Some bacteria have a very definite form, like the *Streptococci*; others, including most marine bacteria, vary considerably. Because of this characteristic, marine bacteria are very difficult to culture and identify, due to changes of form, and it was necessary in my work to observe them directly under the microscope in their natural marine medium. Many varying forms proved to be stages in the development of one type of microbe; they might begin as spirilla, became vibrios, round up into large balls and then break up into small cocci and start all over again, in artificial culture media they became straight rods, quite different from their natural form. This, of course, involved the usual taxonomic difficulties. Marine bacteria are, indeed much more given to changes of form (polymorphism) than even the soil bacteria. Bacteria known to medical science are usually constant in their shape.

Certain groups of bacteria have the power to produce spores. These are round or cylindrical bodies, usually formed within the bacterial cell, which have hard and impermeable cell-walls so that they can stand rigorous conditions, survive where the vegetative forms were killed. They can withstand high temperatures, even boiling water for a few minutes, greater pressure, greater salinity and even desiccation. The shape and position of the spores, if present, are used in classification, some being spherical, some cylindrical, some formed in the middle of the cell, and some at the end – 'terminal spores'.

Some bacteria possess flagella; others do not. These are tiny threads of protoplasm and are used by the bacteria for locomotion. There may be one or two, or many. They too are occasionally lost in culture.

There are a limited number of characteristics that we can use to classify bacteria if we confine our attention to the form that they take. Bacteria are also identified by the effects they have on substances included in culture media. Many different media are used to test their growth powers and fermentative activity; some use common substances as testing materials such as various sugars, milk, potato, starch, paper (cellulose), and so on. Serological tests, which type them against immune blood sera, are an excellent way of identifying pathogenic bacteria, but may not be applicable to soil or marine bacteria. By these means variations in bacteria been found that it has been necessary to use computers to group them in logical order.

As with morphology, these standard criteria often prove invalid when applied to marine bacteria. I have found repeatedly that bacteria which are capable of digesting cellulose, and are therefore important in the destruction of cordage (rope), will only do so if no other, more palatable substances are present. The same applies to other substances such as agar, gelatine and starch. Instead of genera and species, there appears to be a bacterial pool, with groups capable of decomposing the whole range of organic material. One well-known freshwater bacteriologist said that bacteria were transformers and not consumers, and this is generally true. Their main role in the food web is to take complex organic materials and convert them into simpler substances available to other plants and animals as well as for their own needs. This kind of nutrition is known as 'saprophytic' or 'saprobic', and depends on the existence of dead material. Accordingly, one must disregard Linnaean taxonomy for these bacteria, and study them by their products. One must isolate bacteria from an environment, and then recombine them in groups of two or more in the original environment after sterilization. By this means one can get some idea of the range of chemical reactions different groups can mediate.

In addition to these work-horse kinds of bacteria, there are those which only perform certain functions, and even then only under certain conditions. The anaerobes will function only if the supply of oxygen is limited or absent. Some of the anaerobes complete the work of the saprophytes by taking substances containing oxgyen and replacing the oxygen with hydrogen. Thus, carbon dioxide is reduced to methane, nitrate or nitrate to ammonia, and so on. Others, known as chemo-autotrophs, are more strict in their diet, and use only inorganic substances such as sulphates, or hydrogen sulphide for energy. These groups are easy to identify, and interesting, both because of their food requirements and because they are of great antiquity and important in earth history. They resemble most marine bacteria in being Gram-negative and variable in shape.

An interesting group of bacteria which occur in some terrestrial environments are the luminous bacteria. They are more common in the marine world, as they seem to require a fairly high salt content in their environment. They are often to be seen in the dark on decaying fish, but their luminosity is too feeble to be seen in the light. Some of the fish which live in deeper waters where there is little or no light have special organs in which they cultivate luminous bacteria. These organs are glands usually close to the fishes' eyes, and the glands contain a good culture medium for bacteria. The fish have a mechanism for turning the light on and off, one being a blind that they can draw over the gland, and another a means of turning the gland over. The organ is used for recognizing mates, attracting prey, and possibly for lighting up the scene so that the fish can see its prey. The chemical means of producing light in marine organisms including the bacteria is essentially the same as that used by the firefly and the glow-worm. Luminous bacteria are similar in form to other marine bacteria, being Gram-negative and motile, and when they lose their luminosity in culture, which they frequently do, cannot be distinguished from other marine bacteria.

Dr. Robert MacLeod in Canada has shown that certain bacteria require some substances that are found in sufficient quantities only in the sea, while others are inhibited by the same substances in the same quantities.[2]. However, there are contradictions, and one cannot define a series of substances that a typical marine bacterium will require in the concentrations in which they occur in seawater. It might well be that certain bacteria are adapted to seawater conditions, others to fresh-water, while the most versatile can live happily in both.

The typical marine bacteria are pleomorphic (variable in shape), Gram-negative, motile and capable of great versatility in their way of life, except for the specialist groups. Freshwater bacteria are also mostly Gram-negative, but not nearly so pleomorphic as the marine. They include bacteria which are washed into the marine environment in sewage – some of which are pathogenic or potentially pathogenic; luckily, those that are dangerous are rapidly killed by seawater. They are too specialized to withstand marine conditions. So the sea can be said to purify sewage in time.

I found when studying the bacteriology of inshore fishes that there were many Gram-positive species concerned, especially soon after the fish had been caught. There were cocci and rods as well as typical marine forms such as had been described by ZoBell. In the sharks, as they spoiled in the market or the fish shop, most of the bacteria were Gram-positive all the time, and were harmless members of the diphtheria group. Similar bacteria were at times quite frequent in estuarine sediments. Sharks and sting-rays contain a large amount of the nitrogenous substance urea, and these bacteria could split urea and use it as a source of energy, a specialised environment within the marine one. There is another special environment in the sea, namely the intestines of sea-mammals such as the whale, the porpoise and the sea lion. They harbour the same kinds of bacteria that we find in human stomachs and intestines, which must be directly transmitted from one animal to another at birth. In ordinary fish, the gut is sterile unless they have been feeding; after feeding, the bacteria are those acquired from the food. Fish can and do digest bacteria; the bacteria that cause the deterioration or 'spoilage' of fish are found on the skin or in the gills, and require oxygen.[3]

While there are many bacterial diseases of fish recorded from fresh water, there are few from the sea. Diseased fish fail to keep up with the rest of the school and do not survive as a source of infection; and so bacteria are apparently not as numerous in the sea as one would expect from the amount of material that is present to be degraded, and is in fact degraded. We may have underestimated their numbers due to our rather primitive methods, but microbes other than bacteria also assist in the work of destroying dead matter in the seas. The maximum numbers occur where we would expect them, normally about 50 to 100 metres down, where the phytoplankton is most numerous; and occasionally at the surface, where we also find large plankton concentrations. The bacterial concentrations taper off below; but no one seems to have studied their numbers in the laboratory at the pressures and temperatures at which samples were taken. We have all used an arbitrary temperature, (around 22°C) and the laboratory pressure. There are also concentrations of bacteria at the boundary between water masses. Whatever our findings may be, it is certain that there are sufficient bacteria in any area to produce quickly a population necessary to deal with all situations such as plankton swarms or blooms, and that these bacteria are provided with a full range of enzymes.[4] The garbage disposal service is always efficient, and little if any finds its way to the bottom of the deep seas. Forty per cent of the floor of the world oceans consists merely of a red clay with little organic matter.

Viruses are not as common in the sea as they are in terrestrial environments, but some pathogenic viruses and bacteriophages have been reported. The latter are viruses which attack bacteria, and are important in controlling bacterial diseases. There is evidence that the viruses that are found are specific for marine organisms. Viruses are classified according to their hosts, and are probably more widespread in the seas than we realize, since little work has been done on them.

The role of fungi in the sea is not well known, since most of the work has been done on species which are important to wharfage and ropes, dock structures, fishing gear and mooring-lines. As these materials are not natural marine products, the study contributes little to our knowledge of the fungal processes in the oceans or estuaries. Some of these fungi seem to have developed characters which allow them to adapt specifically to the estuarine environment, but there is little evidence to show whether these are important in the ecology, or merely interesting inhabitants of the environment.[5] It has been found that in fresh water, a number of fungi can and do parasitize plants and animals, and it would be thought that a similar situation would occur in the seas. However, there is so far no evidence that this is the case. The phycomycetes, the group responsible for parasitism in fresh water, have been repeatedly demonstrated in seawater, but it still has to be shown that they are important.

A limiting factor in the distribution of fungi in marine environments is that they all require oxygen for respiration, except for certain yeasts like brewer's yeast. This means that they could occur only in the water and on the surface of the sediments; if they do degrade organic materials, they could only do so to a limited degree and not as completely as the bacteria.

Apparently, all groups of fungi are present in the seas, but only primitive members of these groups. Yeasts are present, and may be somewhat prevalent in seawater, though not in the same numbers as bacteria; they occur in all the oceans from the tropics to the Antarctic, apparently in relation to ocean currents. The most frequent of the yeasts is one which has been shown to be a stage in a primitive 'smut' fungus, though we have yet to find the smut stage actually growing in the sea. On land the smuts are most prominent on grasses such as wheat and rye, and on other higher plants which do not occur in the ocean; they do not occur, as far as is known, in the sea-grasses. If fungi were important as parasites, they could have a strong bearing on the timing and duration of plankton swarms and blooms. It has been found that in fresh water, parasites do not inhibit phytoplankton blooms, but merely delay them, and this could allow an unparasitized species to take over from the parasitized one. So it is important to discover whether the fungi do act indirectly in the seas to control the succession of plants and animals in the plankton.

## NOTES

1. ZoBell, C. E., *Marine Microbiology.* Cronica Botany Co., Waltham, Mass: 1-246,
2. Macleod, R., 1965. 'The Question of the existence of specific marine bacteria', *Bacterial Rev.* 29: 9-23.
3. Cutting, C. L. 1955. *Fish Saving.* Leonard Hill, London.
4. Sorokin, Y. I. 1971. 'Bacterial populations as components of oceanic systems.' *Mar. Biol. 11,* (2): 101-105.
   Standon, E., and Parsons, T. 1966. Small particles in seawater; *Lim. Oceanogr. 66, 12,* 367-75.
5. Gessner, R. V., Goos, R.D., and Sieburth, J. McN. 1972. 'The Fungal microcosm of the internodes of *spartina.*' *Mar. Biol. 16,* (4): 269-73.

# 5. THE PHYTOPLANKTON

The microscopic plants of the seas may be divided into three main groups, each of which has distinctive characters and a somewhat different rôle. These are: 1) the diatoms (see below) which were believed to constitute the main element of the phytoplankton; 2) the flagellates (see p.55) which were formerly regarded as of much lesser importance, and which are regarded as plants by zoologists, and as animals by botanists; and 3) the blue-green algae, (see p.48) which may or may not have a blue-green colour, but have some strange and little-known functions, such as fixing dissolved nitrogen and turning it into organic nitrogen and ammonia, a process with a high energy requirement.

The plant life of the seas is as difficult to classify satisfactorily as the bacteria.

## Diatoms

The otherwise unremarkable cells of diatoms are protected by an external skeleton made of silica. This skeleton is very complex, and frequently shows beautiful patterns under the microscope. For this reason diatoms were studied, especially in the early days from about 1800, mostly by amateur microscopists. The amateurs delighted, for example, in arranging the diatom skeletons (or 'frustules') in geometric patterns on microscope slides, using human hairs to push them into place. Some of these fascinating slides with twenty or thirty frustules arranged in concentric, cruciform or triangular patterns are still in

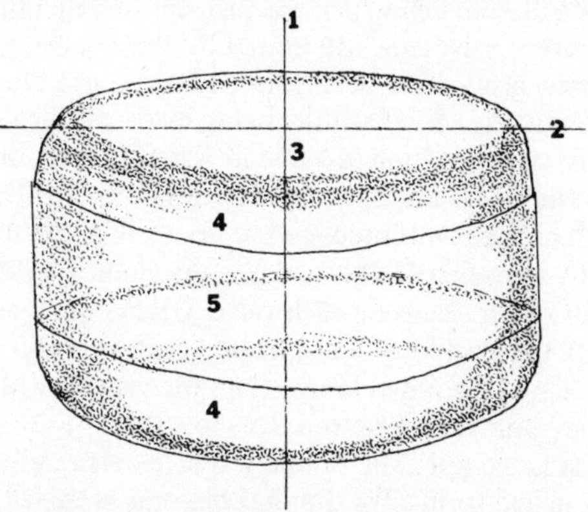


Fig. 9. Parts of a diatom. 1. perivalvar axis, 2. longitudinal axis, 3. and 4. valves, 3 being the valve surface, 4 the "mantle", 5. girdle

existence, a reminder of a leisurely and gracious age. They collected and named about 100,000 species, usually after one another. The number of valid diatom species recognized today is about 10,000. Even this number is far too high, and will no doubt be reduced as variations are determined by culture. One unialgal culture of a *Coscinodiscus* (see p.46), a planktonic diatom, yielded five recognised 'species', which were shown to be stages in the life history of a single species. Some species or strains show great variation in form, while others remain constant.

The diatom frustule (Fig.9) consists essentially of three parts, two 'valves' and a 'girdle'. It is like a tiny box, the valves being the top and bottom, and the girdle the sides. It is the valves which have the patterns used in classification. In the Antarctic and in the tropics, frustules have been found which have the pattern and structure of two different 'families' in the two valves, and this casts doubts on the validity of present classification methods. This may not be important taxonomically, but important is to the ecologist, who wants to know what is going on in the sea, what factors are producing these changes, and what these changes mean. A change in the form of an organism may mean a change in its body chemistry, and this may well be induced by a change in factors external to the organism, such as its habitat. Moreover, in the geological record a number of changes in morphology have occurred, and diatoms with easily preserved skeletons are permanent indicators of that change. The interpretation of such changes in terms of a changing environment is therefore of prime importance also to the geologist.

When diatoms divide, they do so in a direction parallel to the valves, and, as one valve (the daughter valve) is smaller than the other, the cells diminish progressively in size with successive divisions. If you look under the microscope at a diatom cell which has just finished dividing, you will see that the two parent valves are still united by the original girdle, but new valves and a new girdle have been formed inside, and this is why there is a reduction in size. The frustule, being made of silica and rigid, cannot grow. Thus, size is of limited value in classification. In many species of diatom the valves are flat, slightly convex or slightly concave, but in others, such as the planktonic *Rhizosolenia,* they are drawn out and conical, usually asymmetrically so, while the genus *Biddulphia* has two or more knobs or processes on each valve. Others, such as *Chaetoceros* (see p.46) have long, hair-like processes extending from the valve, and these may be much longer than the valve is wide. Sometimes processes may join two or more valves together, forming chains of cells; such chains are characteristic of many species, especially of planktonic and attached forms. We thus get diatoms arranged as single cells or two recently divided cells remaining united, chains of cells attached by processes of varying design and size, and stacks of cells in sheets or ribbons attached by the valve surfaces or short processes extending therefrom. The attachments are all from the valves and not from the girdle. The girdle may be simple or complicated, consisting of a single band joining the valves or a series of scales or plates which may be circular or scale-like, and may be smooth or sculptured. As a general rule, the girdle does not have the complicated sculpture that the valve possesses.

This valve sculpture, used so much in classification, consists of pits and ridges, sometimes very fine, or quite coarse, sometimes symmetrical, or random. Under the electron microscope, it can be seen that there are chambers in the valve and that what we are seeing under the light microscope is the outline of these chambers. With the scanning electron microscope, the view of the frustule becomes three-dimensional, and the valve markings seem more complex. A diatom frustule may be from about 5 to 200 thousandths of a millimetre (one species is about 1 millimetre but it is an exception) in length, yet the complexity and regularity of the valve surface is perfect. The largest diatom is called *Coscinodiscus rex*. It is a cylinder about 1 millimetre in diameter and a millimetre in height with a somewhat dome-shaped valve. It seems to have occurred in warmer waters, and is found on the sea bottom in abyssal trenches, sometimes in large numbers. In the antarctic, there are several species which are a millimetre long, but very narrow – about 1/100 mm. In warmer waters though, many species are up to half a millimetre long; there are also some very minute forms which have only recently been discovered, many not yet described. They can just be seen under 1,500 times magnification, and no structure can be recognised under the light microscope. Sometimes they are very numerous, and they may well play an important part in the phytoplankton. Very small diatoms also occur in the sediments, but have not been described because they are usually stuck to sand grains and other particles, and cannot be separated without disintegrating.

Diatoms, as stated, decrease in size as they divide by fission, without the sexual process. However, a sexual reproduction does not continue indefinitely. When the diminution in size reaches a certain point, depending on the species, diatoms form a large cell or megaspore and this forms a large diatom of the original size, thus restoring the situation. There is also a sexual process in which a number of small, motile cells are formed within the parent. In some species male and female cells can be identified, but in others all the cells are similar. They are motile, swim towards each other and two units to form a large spore, and from this a normal diatom cell. Many species, including most planktonic ones, also form what are known as 'resting spores', cells which differ in shape from the parents, and can only be recognised if we see them forming inside the parent cell. The genus *Chaetoceros,* for example, which has cylindrical cells, forms spores which look something like pincushions complete with various kinds of pins. They have thick walls, usually dark brown under the microscope, can withstand poor conditions, and turn into ordinary diatom cells when conditions are favourable. They are common in sediments, especially on the continental shelf and in estuaries, and they are found in fossil sediments. They act as a reservoir for the species and, under appropriate environmental conditions, turn into ordinary cells simultaneously, giving a sudden boost in numbers. Blooms of these diatoms start on the surface of the sediment, rising into the water as the cells divide and increase. It is usually believed that truly oceanic species do not form these spores, but this may not be true.

There are two main groups of diatoms; one has a more or less circular valve; the other a more or less elongate or boat-shaped one. The latter tend to have a slit or 'raphe' as it is called, and this slit allows the cell protoplasm to have extended contact with the outside world. Such diatoms can move in the direction of the raphe by the extrusion of protoplasm to make contact with a surface. This motion is quite rapid considering the size of the microbe.

It was believed that only the diatoms with a raphe are capable of movement. However, I have seen definite movement of two kinds among the circular (or 'centric') diatoms. With *Coscinodiscus* the disk-shaped form, the motion was intermittent and the disk appeared to jump and twist as though by a sort of jet propulsion. These forms usually have a series of tiny pores in the valve, and it may well be that water is squirted out through these to give the required motion. The movement is limited and not of the rapid kind we see in diatoms with a raphe. On the other hand, I doubt if the latter could move in a liquid medium without a solid substrate; in fact, most of these forms are associated with sediments or are attached to surfaces. Another kind of movement is a sort of swaying, rotary motion seen in another centric diatom, this time a cylindrical one called *Rhizosolenia.* This again is a motion requiring a liquid medium, but I am at a loss to explain its mode of action. Nor do I know how frequent these two types of motility are in nature. I have rarely seen them in collected samples. The motility of the elongate or boat-shaped ones may give rise to very interesting phenomena, as they can move quite rapidly for a limited distance. I have already mentioned the vertical movement of *Hantzschia.* It is hard to imagine the amount of energy required for these small cells to force their way up between the sand grains and then down again. Of course, they gain by being able to orientate themselves so that they are always in the best position for maximum photosynthesis. It is hard to understand why this phenomenon occurs in only one species when many others have the same capacity for mobility, and, under most conditions, occur in the same habitat.

There are two other species which show a form of motility which seems to be useless to the organism. One is a diatom which was described about two hundred years ago called variously *Bacillaria paxillifer, B. paradoxa* and *Nitzschia paradoxa.* It may be seen in a sort of palisade formation with all the cells side by side. Then the individuals glide past each other and the group forms a series of Vs and Ws, each cell retaining contact with the next for a part of its length. They move and sway rapidly, altering their relative positions, returning from time to time to the palisade formation. The other species is *Nitzschia seriata.* It moves in somewhat the same way, but not to the same extent and does not take up the exaggerated Vs and Ws. As these species contain chlorophyll, and the last-named is planktonic, there seems to be no advantage in the motility unless to get more nutrients in contact with the cell.

Many species, mainly of the boat-shaped diatoms, form ribbons or chains of various shapes by means of gelatinous pads, which they excrete, often from the ends of the cells. Many are attached to the substrate by this pad, which may be quite long – as long as the cell, but only produced by the first diatom of the group. Others join at their bases to form a fan or series of fans, or zig-zags, according to whether the pads are formed at one end or both ends of the cell. One genus, *Licmophora,* forms quite complicated and beautiful fans, and the shape of another grouping is suggested by the name *Grammatophora.*

The planktonic species have usually some method of remaining in the photic zone although they are slightly heavier than seawater. Some possess hairs or 'setae', such as *Chaetoceros* and *Bacteriastrum.* Other diatoms obtain buoyancy from oil they secrete as a storage product, this being lighter than water. It is also possible that they store minute gas bubbles in the chambers of the frustule. It used to be believed that most centric diatoms were planktonic, and the boat-shaped or 'naviculate' diatoms were attached or benthic. However, both groups appear in all the habitats.

There is an interesting gradation in shape from the disk-shaped *Coscinodiscus* to the very oddly-shaped *Campylodiscus,* through the naviculoid species. *Chaetoceros* is oval, while *Biddulphia* may be oval, elongate or even triangular in valve view. These are still called centric forms. The genus *Rhabdonema,* shaped like a cigarette box, with the valve a long narrow lid and is placed with the naviculate (pennate) ones, though in reproduction it is allied to the former. There is a series of genera with elongate valves, often with parallel sides, but still without a raphe. Then the raphe appears on only one valve in a few attached genera, one of which is found on whales – the minute carried by the gigantic. The next group in the arrangement has the raphe on both valves and in the centre of each valve, but later forms have the raphe at the side of the valve, and the valve twists into some bizarre shapes, sometimes appearing like a set of dentures, as in some species of *Campylodiscus.* There does seem to be a gradation of form which could mean a line of evolution. However, the diatoms are at the end of one of nature's experimental lines, and taxonomists are not quite sure where they should be placed in the major classification.

### The Blue-Green Algae

The blue-green *Trichodesmium* produces tropical red tides and fixes nitrogen, but differs little in its form from another, *Oscillatoria*, which does neither of these things. This is the only truly planktonic blue-green. 'Nigger heads' too, those black-covered corals, may be due to blue-greens belonging to a group, the Nostocaceae, which are known to fix nitrogen and may well be important in nitrogen fertilization of coral reefs. Some of the species which are found in the estuaries can fix nitrogen, but many cannot. It is believed by some that blue-green algae occurring on substrates such as sunken ships, and on man-made structures below the sea, produce the toxins that appear in fish and cause 'ciguatera' disease responsible for the death of people who eat them in tropical regions. Some blue-greens do produce toxins, especially the genus *Microcystis,* but there is only circumstantial evidence for the association of this or any other genus with ciguatera.

These algae have no nucleus, and very little morphology that one can use for classification, except for overall shape and the presence, absence, or complexity of the gelatinous sheaths that surround the cells. Even the chlorophyll is not contained in chloroplasts as it is in all higher plants. There are some filamentous bacteria that are associated with the sulphur cycle in fresh and salt water, and these, apart from being colourless, are very similar in form to some blue-greens. This is variously interpreted as showing that the blue-green algae are really bacteria, and that the bacteria grade into the algae through the blue-greens. Blue-greens, as we have already seen, are often associated with sulphurous environments, especially sulphur springs.

In addition to the gelatinous sheaths which these algae produce, they may also have sheaths of calcium carbonate; they may, indeed, build up quite extensive lime deposits, greater in weight than the algae themselves.

### The Dinoflagellates

Dinoflagellates are the most important group of microbes in the phytoplankton, apart perhaps from the diatoms; they greatly outnumber other groups in oceanic situations especially in the tropics. Inshore, however, the diatoms are usually more numerous and represented by more species; the same applies in cold waters in both arctic and antarctic regions.

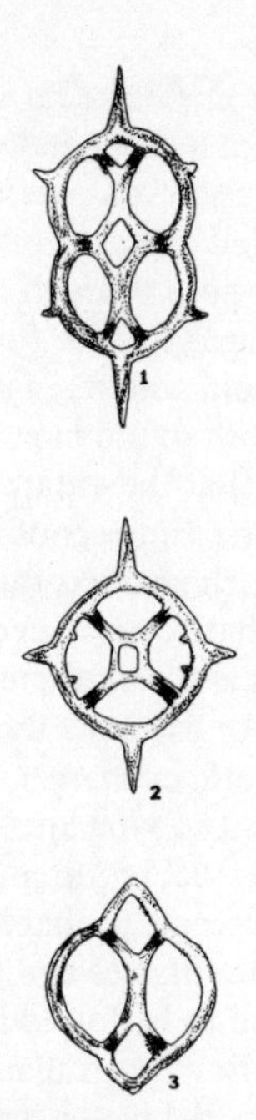


Fig. 10. Silicoflagellates. 1. *Distephanus,* 2. *Dictyocha,* 3. *Mesocena.*

The dinoflagellates are divided into two sections – the 'naked' and the 'armoured'. The naked ones tend to have a fairly tough cell wall, but some are easily destroyed by preservatives, being variable in this respect. Formalin-preserved samples are not suitable for study because of distortion, and a much wider variety of species can be identified when fresh material is studied. The armoured dinoflagellates, on the other hand, have a cell wall enclosed by a series of plates made of a cellulose-like material. These skeletons, while not as long-lived or as characteristic as those of the siliceous diatoms and silicoflagellates (see Fig.10) or the calcareous coccolithophores, are often preserved in the fossil record. Some are of the greatest beauty and complexity, and it is hard to realise that they are merely a single cell with numerous appendages and odd spatial arrangements. The most beautiful belong to the genus *Histioneis,* which occurs only in the tropics, presumably in deeper waters, for it is doubtful whether the genus has chlorophyll. These forms have a series of open channels, which, when they move forward, direct the water down across a groove called the 'sulcus'; by this means particles can be admitted to the cell, so the whole arrangement may be regarded as part of a feeding mechanism. Other forms such as *Ceratocorys horrida* and some of the Ceratia are quite bizarre, in particular the goose-stepping *Dinophysis miles.*

The dinoflagellates, like the diatoms, have far too many listed species and genera. Cultures derived from a single individual may show changes in form covering several species and even overlapping genera. The species involved in the Gulf of Mexico red tides, *Gymnodinium brevis,* also has a very large variation in form as it goes through its life history.

Most of the species of dinoflagellates are found in tropical waters. Only a few are found in really cold waters, mainly belonging to the genus *Peridinium,* which resembles those found in estuaries in warmer waters, thus supporting the contention that the antarctic microscopic plants indicate estuarine affiliations. The same genus, *Peridinium,* occurs frequently in the North Sea, which also has the character of an estuary.

It has long been contended that certain species of phytoplankton are restricted to certain water types and can, therefore, act as 'indicators' of these types. It has been shown for example that the diatom *Rhizosolenia curvata* occurs only in the antarctic convergence, within the temperature-salinity limits of that mixing area. There has been no detailed documentation of the actual requirements of most species that are used as indicators, though it seems probable that species do indicate certain limits with regard to water masses and their movement. For example, a number of species are to be found in equatorial waters and carried by currents, such as the East Australian Current and the Florida Current, into temperate areas. As the water temperature drops, many of the species disappear in turn, though they have been carried along in the same water mass. They seem to show the same pattern in the western Atlantic and the south-west Pacific, so may indicate similar conditions in the several oceans. This is important to the ecologist, who must rely on a realistic taxonomy closely related to ecology rather than to a single type specimen.

Of the dinoflagellates, *Peridinium* is mainly estuarine; species found in the open ocean probably originated in inshore situations, but are tolerant of true oceanic conditions. *Ceratium*, on the other hand, is largely oceanic, and only a few species are to be found living permanently in the estuaries, to which they are rarely confined. A few of these are cosmopolitan and occur from the sub-Antarctic through the tropics to the Arctic. Two or three species of *Ceratium* occur in fresh water, and in the Florida Everglades, but not in Biscayne Bay or the Straits of Florida; they seem restricted in their salinity tolerance.

A group of armoured dinoflagellates, the Dinophysidae, has two plates on each side of the cell and a circular plate joining them. It also has various transparent or translucent wing-like appendages, some of which are large and beautiful as in the case of *Histioneis* which belongs in this group. So does *Dinophysis miles,* the goose-stepping species which is found only in the Indonesian region. Others of this group are mostly confined to tropical waters, usually below the photic zone, so the presence of many species of this whole group indicates deep, warm water, or at least sub-surface water derived from the tropics. The group ranges from rather ordinary looking forms with small appendages – these are more cosmopolitan – to bizarre forms, one of which is very much elongated (Amphisolenia) and another has a central body with three extensions (*Triposolenia*). Another genus, related in appearance to both *Peridinium* and *Ceratium*, and called *Heterodinium*, is found in sub-surface waters, has apparently no chlorophyll, and is an indicator of warm water upwelling from the aphotic zone. Thus, the dinoflagellates are useful as indicators of both horizontal and vertical water movements.

Dinoflagellates are distinguished from the other flagellates by having one flagellum which trails behind and propels the microbe forward and one which is coiled like a spring and lies in a groove around the organism. This groove is called the 'girdle' but is different in appearance and function from the girdle of the diatoms. The two flagella give the organism a wobbly forward motion, of a rather drunken character, which is quite characteristic. Both flagella arise in a groove running fore and aft particularly in the after-part of the cell, this groove being known as the 'sulcus'. The motion of the organism and of the flagella tends to create a flow past the sulcus so that particulate food can be taken in there, as in the 'gullet' of other forms with which the sulcus is homologous.

The naked dinoflagellates vary in size, from quite large ones to very tiny ones a few thousandths of a millimetre long. They occur in seawater and in the sediments, particularly in sands, and are probably very important as food for many small marine animals. Some of them have been photographed in the act of cannibalizing each other. In warmer waters in particular, they are at times very abundant and may be the dominant species, quite apart from their occurrence in red tides, both toxic and non-toxic.

An interesting group in the plankton is the genus *Pyrocystis*. This genus consists of quite large forms, and the cells are rather like lumps of jelly, spherical, crescent-shaped or biconical, and visible to the naked eye. Only when they reproduce do they resemble dinoflagellates, and then a series of spores are formed which look like *Gymnodinium*. They break out from the parent mass, swim about, and eventually turn into the large plants that we can see in the phytoplankton. Another, even larger organism, which does not in itself possess chlorophyll, is *Noctiluca* (translation of the name – night light). It normally has no resemblance whatever to a dinoflagellate. It is luminous, and, in Indian waters, it contains within it a sort of imprisoned plant cell, or zooxanthellae, which is taxonomically unrelated to it.

**The Coccolithophores**

Another group of flagellates which are important in the fossil record, but which nowadays seem largely confined to warmer waters, are the coccolithophores. The name in Greek means 'round stone carriers'. They consist of small cells, usually less than 30 microns in length, round or elongate, containing chlorophyll and other pigments to make them yellow-green, and having embedded in their cell wall a number of calcareous pieces called coccoliths. Coccoliths are marvellously symmetrical bits of limestone shaped like collars, spines, clubs and disks, each shape characteristic of one particular microbe. Often there is an internal structure in each coccolith that you can see only under the electron microscope. There are probably up to two hundred coccoliths around each coccolithophore. We must assume that they are able to live in the dark, either feeding on dissolved organic matter like the bacteria or small particles like the dinoflagellates.

Coccolithophores are very common in certain waters, particularly in some tropical waters, and have some value in indicating the origin of certain water masses. They are not common in the warmer waters of the southwest Pacific or the eastern Indian Ocean, but are often abundant in Timor Straits; one species, *Discosphaera thomsoni,* with flower-like coccoliths, is abundant in the Brazil Current and waters derived therefrom as they go into the Caribbean and even through Yucatan Channel into the Straits of Florida. However, the coccolithophores as a group are much more important in the Tongue of the Ocean, in the Bahamas, than in the rest of the western Atlantic. They are of major importance in the waters of the Benguela and Guinea Currents, and in the Mediterranean.

The group also appears to be of great use as food for plant-feeding animals of the zooplankton. I had evidence of this in a study we made on the equator in the Gulf of Guinea, south of Nigeria. During the day, the number of coccolithophores increased to millions per litre. As dusk fell, the numbers dropped to tens of thousands, which seemed to indicate that they were being eaten. The drop in numbers coincided almost exactly with the upward movement of the zooplankton which, in many places, occurs in the evening. The next day, after sunrise, the coccolithophores reproduced until, that evening, the numbers were as great as before. In this case, the upward movement of the zooplankton was observed on a precision depth recorder (PDR) which can measure depth to the nearest metre, and can also detect solid objects or layers such as fish or plankton swarms. In certain waters such as the Gulf of Guinea, the animals live in a belt at certain depths, the 'deep scattering layer' or DSL.[1]

Parts of the DSL move upwards towards dusk, presumably as part of a feeding movement. The animals seem to come towards the surface in batches, presumably of different groups or species. They tend to stop at sub-surface levels, and these levels are usually those at which most of the phytoplankton occurs. The assumption that this is a feeding migration seems obvious, especially as the number of microbes in the phytoplankton decreases at that time. In this case, the coccolithorphes and the diatom *Coscinodiscus* diminished greatly at dusk, but *Rhizosolenia* did not. It has been suggested that *Rhizosolenia* is not acceptable to certain zooplankton.

The restoration of the coccolithophore population within 24 hours can only be interpreted as occurring by reproduction, and this at a rate far in excess of that calculated for any phytoplankton from culture experiments. Division must have occurred every 2 hours or less to give the numbers that we found.

The plants apparently multiplied fast during the day and were eaten by the zooplankton of the DSL during the evening migration. After that, when the second and third waves of zooplankton made their vertical migration, feeding on the plants diminished or ceased, due either to diminishing returns, poor light or to satiety. The later waves may well have been feeding on the earlier ones and not on the plants at all. Lastly, there were the predators who waited all night at the original DSL level and consumed the well-fed migrants on their return to lower levels.

**Silicoflagellates**

Silicoflagellates (Fig.10) appears to be much more numerous in fossil materials than they are today; there is some doubt about this, since once the microbe has died the heavy silica sinks rapidly to the bottom of a shallow estuary or sea.

There are several silicoflagellates to be found in the phytoplankton, but they rarely form an important fraction. Some have complex siliceous skeletons, like *Distephanus*, others are simple. They are usually present in antarctic samples, but occur in warm water also. On one occasion, in Peril Strait and Sitka Strait, I found a bloom of these flagellates, the only time this has been observed.

**Other Flagellates**

I have implied that the importance of the flagellates in the phytoplankton, and in the marine environment generally, has been greatly underestimated, because phytoplankton was, until fairly recently, collected in nets not fine enough to catch the smaller microbes, and secondly because of the sensitivity of many flagellates to surfaces and to all preservatives. Many species will burst if they touch a surface such as a glass slide or a filter, and many more will appear as blobs of jelly if you add formalin or even iodine (another usual preservative) to them. These are the ones that have no protective cell wall, and even some with internal skeletons such as *Distephanus* disintegrate very readily.

A number of groups of flagellates such as the Xanthomonads and the Chrysomonads are quite common in the sea. Rather less so are the Chlorophyceae and the Euglenids.

The use of the term 'phytoplankton' to denote the plant components of the total plankton is unfortunate, because so many of the organisms that are usually included may or may not be behaving as true plants. Moreover, there are important plant-animal associations which are not usually included with the phytoplankton, but do contribute, sometimes very substantially, to the productivity of the area. The word 'protoplankton', meaning 'earliest plankton', is used in the sense of the most primitive plankton. There is also a very close association between the organisms which do behave as plants and those which do not, so much so that some individual cells can and do revert from plant to animal nutrition. I refer here specifically to the protozoan *Euglena* and its relatives. *Euglena* has a primitive 'gullet' and can ingest and digest other small microbes, but also has chlorophyll and can behave as a true plant. However, it can easily be made to lose its chlorophyll and live on dissolved organic matter and particles only. When treated in some ways, it can regain its chlorophyll, but in other ways, not, and it must then remain as an animal. As we have seen, other flagellates have this dual type of existence, but *Euglena* has been the subject of intensive study and is an excellent experimental microbe.

Another interesting flagellate, closely related to *Euglena,* and with a close colourless relative known as *Astasia,* is *Eutreptia.* It is to be found in estuarine sediments, but is also frequently present, though rarely numerous, in oceanic phytoplankton samples. This form as well as having the gullet and flagella of a *Euglena,* also resembles an amoeba, which can change its shape and roll along like a moving jelly. Living *Eutreptia* is always doing this; it will sometimes appear like a nail with most of the protoplast in the head; then the bulk moves up and down the cell like an apple being shaken up and down in a sock. It rounds up into a ball if any stimulus is applied, so is not recognized in preserved samples. It has been previously found in sediments, but I think mine is the first record from oceanic plankton.

The other groups of flagellates which we have mentioned in the preliminary list are found at times. A few species of the green flagellates (Chlorophyceae) and sometimes quite common, but in limited areas. The yellow flagellates (Chrysophyceae) are rather more widely distributed, and seem to occur in most phytoplankton samples; they are sometimes dominant, but usually in smaller numbers. Only a few species are known, and they seem to be oceanic and inshore, but their true distribution is difficult to estimate, mainly because they are hard to recognise in preserved samples. A tiny flagellate, *Phaeocystis,* forms large blooms in the Antarctic and occasionally in the tropics. We have no information about the reasons for its occurrence. In the Coral Sea, on several occasions, I have found phytoplankton samples which contained nothing but very minute flagellates that were green, but so small that one could find nothing to characterise them; they were about 1/1000 mm in diameter. They also occur in quantity in a belt from South America at least to the Tasman Sea in the southern part of the subtropical convergence. Under the fluorescent microscope they appear as minute red dots or as a red haze. They are certainly dominant, and even the only organism in these areas, so must be important in the overall phytoplankton productivity in some parts of the world. The German planktologist Lohmann described[3] a large number of these minute flagellates caught on the very fine filters of pelagic salps. Such ultra-microscopic organisms constitute the so-called nanoplankton; though few of these have been described, their total volume and importance may be as great as that of all the other phytoplanktons combined.

This has become known as the microplankton or μ-plankton, composed of such organisms as well as the tiny diatoms and dino-flagellates, may be many times as significant in the total mass of phyto-plankton as the larger microbes.

Some of the green flagellates of the Chlorophyceae are very similar in shape and flagellation to spores of larger seaweeds, and may at times be confused with them. In fact, it is now becoming clear that many flagellates are stages in the life history of other organisms; the coccolithophores, for example, seem to be stages in the life histories of other algae, some unicellular and some multicellular. This confuses the issue, but when solved it will tell us more about the evolution of the more primitive species.

**Colourless Forms – Flagellates**

The coloured *Euglena* and its allies have their colourless counterparts, which are heterotrophic, feeding mainly on organic particles. There are other colourless flagellates in the oceans, whose classification has not been attempted. I have found them, at times, extremely abundant following phytoplankton blooms, and accompanying dead phytoplankton cells on which they appear to be feeding avidly. They seem to be utilizing the detritus, and are probably also a good source of food for the higher animals; but because the main scientific interest has been in the producers, i.e. the chlorophyll-bearers, little is known about their importance. Like the coloured flagellates, they are easily destroyed by preservatives and do not survive confinement in small vessels; so it is not easy to study them, except in freshly collected samples. Moreover, they move around fast and are not easy to observe unless you can slow them down in some way. Apart from some of the colourless dinoflagellates, they have no hard parts, and so are very impermanent in the geological record.

**Colourless Forms – Ciliates**

I have frequently mentioned the ciliates (See Fig.6, p.31), those protozoa which have a number of rigid, but moveable spines, and often what is known as an undulating membrane, consisting of innumerable hairs which move in a rippling motion in unison. The ciliate *Paramecium*, which is one of this group, is well known. The planktonic ciliates belong mainly to a group known as the tintinnids, which are free-swimming, but live in a small cup called a 'lorica' which they build around themselves out of the materials at hand, some species using sand grains which they stick together and others making a chitinous sheath. The spines or cilia usually project from the top of the cup, but can be retracted if there is danger; they are used both for mobility and to pass food in the currents they set up across the gullet. The tintinnids are found in plankton samples from the equator to the antarctic waters and are very numerous at times. Almost all the work which has been done on them is taxonomic. Closely related is a group which also makes a cup to live in, but attaches this at the end of a long and spring-like stalk through which there runs a protoplasmic, contractile thread. Some species form colonies, with a dozen or so animals at the ends of their stalks, all of which are attached to a common spot at the other end. Each animal will slowly extend its stalk until it is almost straight and then let it spring back simultaneously with the other colonists, in perfect synchrony. How does this animal manage such synchronization while living at the far end of the stalk. These ciliates, of which *Vorticella* is a laboratory example, are found in the ocean attached to floating *Sargassum* and to a chain-forming diatom *Chaetoceros coarctatum.* Again, why is only one diatom species involved with *Vorticella*?

Apart from these sheathed ciliates, some of which have "imprisoned phytoplankton" called zooxanthellae in them, there are a number of naked forms in the plankton and an even greater number in the sediments. The planktonic ones are often extremely abundant with the flagellates in dying phytoplankton and zooplankton swarms, and if they appear in phytoplankton cultures in the laboratory spell the rapid doom of the algae, consuming it entirely and very quickly. In the sediments, they play a very important role, eating bacteria and other microbes and maintaining an ecological balance. Little is known of their use as food for other animals either in the water or the sediments, but they must be consumed.

Certain phytoplankton organisms such as the diatom *Rhizosolenia*, are not eaten directly by the zooplankton. This diatom produces large blooms in the Barents and North Seas. The very important zooplankton animal *Calanus* avoids blooms of *Rhizosolenia*, but does feed around the edges of the blooms and below them. The distribution of flagellates and ciliates in this kind of situation indicates that *Calanus* is feeding on them, but this is unproved. It has been found that different species of ciliates play different roles in the community, some feeding on bacteria and some on microscopic algae, but each being somewhat restricted in its diet.

**Colourless Forms – Foraminifera and Radiolaria**

Not usually classed with the phytoplankton, but often living in association with it, are two other groups of protozoa, the foraminifera and the radiolaria. They are of the greatest interest to geologists, because they have changed in morphology through geological time, and have hard parts which remain in the sediments. The foraminifera (Fig. 8) have pretty and complex shells, usually convoluted and made of calcium carbonate; they can be seen with the naked eye and are usually on display in museums. They are protozoa without pigments so cannot be photosynthetic without zooxanthellae. When these are present, as is usual, they can contribute to primary production and act as part of the phytoplankton; and, being motile, they can carry their zooxanthellae around from place to place. Some foraminifera use strontium instead of calcium to make their shells, and this could have a serious effect on the environment in areas of radioactive fall-out where they might well concentrate the highly dangerous strontium-90.

The radiolaria, like the foraminifera, are related to the amoeba; they have a skeleton usually made of silica and often very beautiful. These animals also contain zooxanthellae so may be regarded as potential primary producers. The importance of these two groups as primary producers is lessened by the fact that they are not restricted to the parts of the ocean where light penetrates, and are apt to occur at somewhat random depths. The plant components may or may not get a chance to photosynthesize profitably, but perhaps it is enough if they do so at all, provided that they supply certain needed substances to the animal.

**NOTES**

1. Schmidt, W., 1972. 'Deep scattering layers'. *Mar. Biol. 16.*
2. Savage, R. E., and Wimpenny, R.S. 1936. 'Phytoplankton and the Herring, Pt. II.' Fish. Invest. London, *15,* No.1.
3. Lohmann, H. 1903. 'Neue Untersuchungen uber den Reichtum des Meeres an Plankton.' Wiss. Meeresunters. Abt. Kiel. *8*: 1-86.

# PART III

# THE PHYSICAL WORLD OF THE OCEAN

# 6. SUNLIGHT AND LIFE

We know that sunlight provides the energy for all living matter, but we probably do not fully realize that it is the only *external* source of energy for the earth as a whole. If it were not for sunlight, the earth would not only be a dead planet, but would have cooled rapidly as it lost its irreplaceable energy. This loss of energy, which accompanies all processes throughout the universe, is known as 'entropy'. Newton's laws state that energy can neither be created nor destroyed, but it can be lost within a system or between systems as in friction between moving bodies. The rate of loss from our earth system is slowed down by the energy of sunlight, chiefly through the agency of chlorophyll. There are other ways in which sunlight can contribute energy which are important in earth history. For instance, finely-divided iron (iron powder), in the presence of sunlight, can catalyze a number of reactions and greatly increase their rate of progress. An old-fashioned method of reproducing plans and drawings was the blue-print process, where an iron salt and sunlight were thc necessary materials. Then, of course, there is the use of sunlight in photography, which is really a chemical process activated and accelerated by light. These photocatalytic processes were certainly very important on primitive earth, and probably played a major role in the evolution of living matter.

The term 'primary productivity' or 'primary production' occurs frequently and readers will probably find also that there are several different definitions of these. Whatever the definition, the process described will depend on the use of sunlight for the energy to combine inorganic or simple organic substances into the complex ones required for living organisms. Primary productivity is the first link in the food-chain, in which organic material is built, and stores the energy obtained from sunlight. It is confined, both on land and in the sea, to green plants, including among these the purple and green bacteria with their rather primitive chlorophyll and somewhat irregular photosynthesis. The term 'primary production' is usually confined to assimilation processes using sunlight as the source of energy, and excluding those other processes in which carbon dioxide is assimilated by microbial use of chemical energy, mainly because the latter processes do not increase the amount of energy in the system. Let me emphasize that all other processes merely use energy that is already present on the earth.

Primary production is possible, therefore, only when light is present. Furthermore, the depth at which the photosynthetic reaction can take place is restricted by the available light, and the penetration of that light into the water. Thus the clearer the water, the greater the penetration, and therefore the depth at which productivity can occur. In clear water this depth is usually considered to be about 450 metres. However, photosynthesis requires light of certain wavelengths, mainly blue and red light, and the seawater filters out different wavelengths at different depths. Further, the water is rarely, if ever, completely clear. Dust particles dancing in a beam of light dance because they reflect light and can thus be seen from one side. This means, of course, that some of the light in the beam is lost due to reflection from the particles, and the more particles, the greater the loss. In a sunbeam, the loss is insignificant, but in the sea, where light is being absorbed rapidly, it can be very important in limiting the depth to which plants can grow. This dancing of particles is known as the 'Tyndall effect', and can be measured either by reflection or by the loss of light from a point of emission to a point of reception. To measure this, one usually uses an instrument which sends a beam from one point to another one or two metres away, and receives it by a photocell so that the loss can be measured. If one varies the colour of the light by means of a series of filters, one can determine the loss of various wavelengths, including the ones that are important to the plant chlorophyll. However, many plants contain what are known as 'accessory pigments', and these, though not capable of photosynthesis themselves, are able to trap light and pass it on to the chlorophyll that does the work. These accessory pigments are the ones that give red, yellow, blue, or other colours to the plants or flowers; they can absorb wavelengths other than those required by the true photosynthetic pigment chlorophyll *a.* In absorbing light, they also absorb some heat and prevent it from damaging the chlorophyll *a.*

Some plants, especially marine ones, can use also a more economical process than the photosynthesis of carbon dioxide. They can use sunlight mechanisms to recombine simple organic substances such as acetic acid (the acid of vinegar) and lactic acid (the acid of milk) in forms suitable for their own purposes. This requires only about a tenth of the energy needed for photosynthesis and therefore can occur at much greater depths in the ocean. Thus, it is now known that the primary productivity or 'photic' zone may be much deeper than was thought a few years ago. If this greater depth extends throughout the oceans of the world, we can assume a much greater possibility of productivity than was calculated previously. In fact, we often find the greatest amount of plant material is at a greater depth than was formerly believed possible. It was formerly assumed that this was dying material that was sinking, but it is in fact quite active and frequently growing.

An interesting discovery made in 1948 by Dr. Francis Bernard of Algiers,[1] was that certain microscopic plants occur in deep waters where there is, in theory, no light at all except the small amount that is produced by luminous organisms that live in the depths of the ocean. Some time later, I was looking at material brought up from the bottom, 20,000 to 30,000 feet below the surface, by the Danish *Galathea* expedition. This was placed by Dr. Claude ZoBell, of the Scripps Institution of Oceanography in California, in steel 'bombs', and kept at a pressure of 1,000 atmospheres (15,000 lb. per square inch) in the dark.[2] After some two years, some of the marine plants (diatoms) were still alive, though a few had burst due to the release of pressure when I opened the bomb. They had no chlorophyll, but then there are diatoms which never contain chlorophyll, such as *Nitzschia putida,* which lives in shallow sediments. Also, these particular ones had been in the dark throughout their lives, and probably for generations.

I had recently read some work of Dr. Joyce Lewin[3], who had grown a number of marine diatoms without light, giving them sugar and glucose instead of carbon dioxide. She had, in effect, induced in them what is called a 'saprophytic' existence similar to that of bacteria. A 'saprophyte' is a 'plant' that lives on decaying organic material and assists in the decay. Many of her species grew under these conditions, and this seemed to explain my findings and those of Dr. Bernard. So using her techniques, I grew the diatoms in the dark, but at a pressure of 500 atmospheres, simulating a depth of 15,000 feet or 5 miles. Under these conditions, they grew quite well and reproduced normally.

The next link in the chain of the research was forged when I made a chance visit to the University of Miami in the summer of 1961. Dr. Eugene Corcoran showed me some cultures that he and his assistants had made from deep water in the Tongue of the Ocean, that almost land-locked basin in the Bahamas. These cultures were made because the assistant, when making cultures from samples taken near the surface, had accidentally included some deep water samples. Although a mistake had been made, Dr. Corcoran fortunately decided to look at the deep samples, which, according to current theories, could not contain living plants. He found that they did, and that in the laboratory they actually behaved as typical plants, and used light and chlorophyll to grow and reproduce. In the cultures were found about 20 species of microbes, all with chlorophyll, and growing well. So far, all studies had been made using methods which did not guarantee sterility, so there could be the criticism that the organisms had been collected by the open sampler on the way down. Dr. ZoBell had designed a sampler which could be relied upon to take a sample just where it was required, so that one could be sure that there was no contamination. I modified some of these for my purpose, and took a number of samples in the southwest Pacific from deep waters to 30,000 feet. In nearly every sample I found microbes that contained chlorophyll, easily demonstrated by its red fluorescence under blue light. Thus it was proven, to my own satisfaction at least, that microscopic plants could live in deep waters which had little or no light, provided that glucose was available. Dr. Mary Parke, working at Plymouth, England, had also grown hundreds of microscopic plants from deep water, but would not publish her results as she was afraid of the criticism that they might be due to contamination. Her results have never been published, but many other research workers have since found confirmation of the observations made by Bernard and myself.

Beyond the fact that it is possible for some of these small plants to live and reproduce in the dark under pressure, we have little evidence as to whether they usually do so, or how important they are. Unfortunately, the techniques are very difficult, and some critics maintain that there is not enough dissolved organic matter in the oceans (substances such as glucose) for the plants to exist; this criticism appears to have foundation. However, another clue may be that many of the deep-water plants are flagellates, which are grouped with the protozoa, and can, at least theoretically, and in many cases practically, absorb and digest solid particles including living ones. It is unexplained also why these microbes waste energy making chlorophyll if they cannot use it: perhaps they just make it in case they or their progency may find some light somewhere. Or can they, by some means as yet unrecognized, use any small amount of light that becomes available in their dark world? It is known now that chlorophyll also assists in the assimilation of nitrogen into the plant, and the chlorophyll may be necessary for this and not for photosynthesis after all.

An exciting sidelight on these deep-water plants came during a voyage I made in the eastern Mediterranean with Dr. Baruch Kimor of the Haifa Technion. He was towing fine plankton nets at set depths to nearly 4,000 metres (12,000 ft.) near the Island of Rhodes. The tows were horizontal, the depths clearly recorded as described in the chapter on methods, and the nets were closed before they were brought up. On examining the fresh material from his tows in the ship's laboratory, he saw some protozoa (radiolaria) which had regularly spaced round cells within them. It is well known that many radiolaria have plant cells which live in 'symbiosis' with them. When examined under a fluorescent microscope there was no doubt that they contained chlorophyll.

One can rationalize plants and animals living together happily and forming a 'commonwealth', but the idea of one partner living in an area where he cannot contribute and must be completely supported by the other 'partner' is hard to accept. Our samples were taken in August, and the nature of the flora and the chemical and physical composition of the water showed that our plants had been there at least since the previous February.

It seems obvious that there is a limit to the depth to which primary productivity, dependent on the penetration of sunlight, can extend. There is probably some production below the average depth, but this must be limited, and in theory there is a depth at which loss of energy, usually measured by the amount of respiration, becomes greater than that gained by photosynthesis and other processes. So the water below the 'photic zone' will be negatively productive.

Light is reduced by being reflected by particles which may consist of the microbes themselves; of inorganic particles of such substances as phosphate of calcium (from bone) and phosphate of iron (a substance so insoluble that it forms a sort of end-particle in the seas); organic particles made up of disintegrating animals and plants, especially their more stable parts such as the calcium phosphate just mentioned, the chitin which forms the shell of lobsters, shrimps and other animals of this important group,sand the agar and alginate which are the structural substances of the red and brown seaweeds. Many of these organic particles stay suspended in the seas for a very long time before they are destroyed by bacteria and other microbes. Some may possibly be re-used without being destroyed, but there are gaps in our knowledge here.

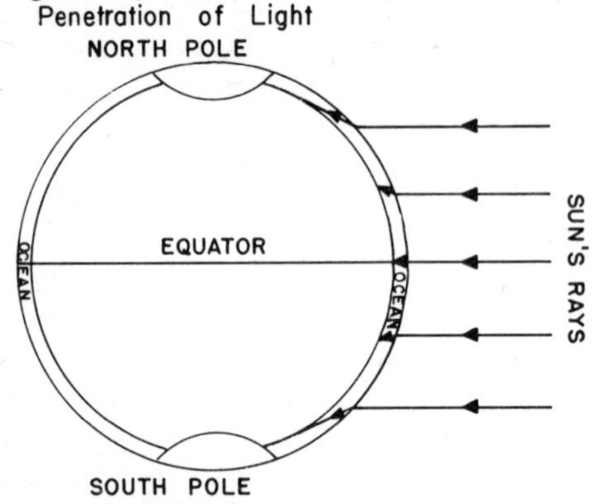


Fig. 11. Penetration of light varies with latitude, being obviously greater at the equator where there is less refraction.

Another way that light penetration is reduced is due to the curvature of the earth (Fig. 11). If the sun is over the equator, light will penetrate vertically without refraction, but once the light ceases to be vertical, it will be bent and moved further from the vertical and so must travel further to reach the same depth. If the surface of the water is smooth, there will be reflection, increasing with the 'angle of incidence' and also increasing with the increase in ripple or wave action. Thus, the depth of light penetration in the arctic regions will be less than at the equator, while penetration will be greatest in the tropics. We find, as we should suspect, that the photic zone is deepest in tropical waters, shallowing towards the poles.

Nevertheless, it is in the tropical waters, or sometimes in the sub-tropics, that we find the greatest concentrations of microscopic plants at the surface. This is surprising, because most of the tiny marine plants tend to avoid high light intensity. Chlorophyll is injured by intense light, and this is why desert plants are greyish in colour; they have heavy cuticles or reflecting layers of non-living material, often beaded to increase reflection, these structures designed to reduce the light penetrating to the chlorophyll below. So the marine plants seek to combat intense light by living normally in deeper water. The presence of animal organisms close to the surface in the tropics and subtropics, and in particular the paradox of what is known as the 'red tide', is explained by their possession of accessory pigments, which protect the chlorophyll.

The earliest records of red tides come from the Middle East, where their existence is recorded in the name given to the *Red Sea*. This sea frequently has a red colour, which is due mainly to tiny microbes known under several names, usually as *Trichodesmium erythraeum*. This is a tiny blue-green algae consisting of bundles of fine threads, and can be found throughout the tropical oceans, being common in the Indonesian region, the tropical Indian Ocean[4] between Colombo and Fremantle, the Coral and North Tasman Seas and other parts of the Pacific, in the tropical Atlantic, Caribbean Sea and the Gulf of Mexico.[5] There are other microbes that produce red tides and these are mainly small flagellates or coloured protozoa; many of these are poisonous, causing the death of fish, for example in the Gulf of Mexico along the west coast of Florida, off Walvis Bay in South Africa and off the coast of Peru where they produce what is known as the 'Callao Painter'[5]. The common property of these large blooms of microbes is that they are generally red or yellowish-red in colour, and not green as would be expected. I have already mentioned the rate of accessory pigments, which feed energy to chlorophyll *a* without being able to use it themselves. They make use of the additional property that they are not so sensitive to high light intensity or to heat, and can absorb much of the excess of these and protect the chlorophyll *a*. In the surface algae these pigments are much more abundant than in the microbes at deeper levels. The red tide microbes depend on them for their existence in an otherwise uninhabitable zone, and as they have a virtual monopoly there, can multiply almost unhindered until they can actually be seen from a ship or an aeroplane. I have seen and recorded a red tide extending inside the Great Barrier Reef of Australia from Gladstone to Cairns, at least 20,000 square miles. It looked like a series of sandbanks from the air, and as I passed through it by ship a few days later I found it to be about 2 feet deep. I have also seen similar quantities in the Coral Sea between the Great Barrier Reef and Noumea, in the Indian Ocean between Colombo and Fremantle, and in the waters of the Arafura Sea.

The *Trichodesmium* red tides do not seem to be toxic. However, those caused by the dinoflagellates *Goniaulax* or *Gymnodinium* may be extremely toxic and cause large fish kills, and may cause an allergy to people walking along the beaches near the red tides. There are *Goniaulax* red tides on the coast of Oregon and Washington, and these exude a toxin which makes the local shellfish poisonous to humans. Red tides with toxic effects are endemic in Walvis Bay, seasonal off Peru and very frequent in the Gulf of Mexico, where they have been the subject of considerable study without any conclusion being drawn as to their real cause. It has been suggested that they are due to the vitamins escaping to the eastern Gulf of Mexico from the farmlands, especially the orange groves of central and southern Florida, to the high phosphate of these waters because of the intensive mining of phosphate rock in the Tampa region, and to physical boundaries being set up by tides and winds between the coastal waters and those of the Gulf of Mexico proper, thus concentrating the algae at the meeting place. Experiments and field data on these supposed causes have been inconclusive. Red tides, especially the poisonous ones, occur in other areas. One was recorded in Sydney Harbour (Port Jackson) at the turn of this century, and one or two have been reported from Port Phillip in Victoria, and occasionally from Vigo Bay and other harbours in Spain and Portugal, but there seems to be no explanation for their occurrences.

Light is the most important single factor in primary productivity on land or sea, but there are many other contributing factors determining the degree of productivity of an area.

## NOTES


1. Bernard, F. 1958. 'Donneés récentes sur la fertilité elementaire en Mediterranée.' Rapp. Cons. Explor. Mer, Copenhagen, *144*: 103-8.
2. ZoBell, C., and Morita, R.Y. 1953. *Bacteria in the Deep Sea. The 'Galathaea' Deep-Sea Expedition.* Geo. Allen & Unwin, London.
3. Lewin, Joyce. 1953. 'Heterotrophy in Diatoms.' *J. Gen. Microbiol. 9*: 305-13.
4. Gunther, E.R. 1936. A Report on Oceanographic Investigation in the Peru Coastal Current. 'Discovery' Repts. XII: 107-276.
5. Galtsoff, P.S. 1948. 'Red Tide'. Spec. Sci. Rept., U.S. Fish & Wildlife Service, *46*: 1-44.

# 7. PRESSURE, TEMPERATURE, SALINITY, DENSITY AND CHEMISTRY OF THE OCEANS

## Pressure

Water is always incompressible, and this is important to microbes. It will not expand very much inside them if they are brought to the surface from deep waters, so they tend not to burst from a slow release of pressure, though a sudden change of pressure may cause rupture. Further, the likelihood of rupture depends very much on the size and complexity of the organism, and on the presence or absence of gases in the cells or organs. The smaller the organism, the easier it is for equilibrium to be adjusted between the interior of the cell or cells and the external world; in addition, the ratio of surface to volume is important since the larger the surface, the more room there is for things to get in and out. This is especially true of gases. We should expect what we find, that single-celled microbes are least affected by sudden pressure change. The hydrostatic pressure in the seas (or elsewhere for that matter) increases about 1 atmosphere (15 lbs per square inch) for every 30 feet of depth. Thus, in the deepest oceans, the pressure will be 1,000 atmospheres or 15,000 pounds per square inch (p.s.i.). It was long thought that nothing could live at such pressures. But by 1820 living creatures had been brought up from a depth of 100 fathoms, and by 1912 a rich fauna of abyssal fishes called invertebrates was known. The Danish research-ship *Galathea* specialised in the abyssal fauna and added much new information. Dr. Braun, leader of the *Galathea* expedition, believed that there might be giant eels in very deep waters, and many others suspected that giant squids and possibly other animals might live in such deep waters. However, because of their strength and weight it would be difficult to catch them with the equipment we have; those that had air bladders or other gas spaces would be killed by the sudden release of pressure on surfacing. This would also happen because dissolved gases tend to escape from liquids when pressure is lowered, as occurs in the familiar bends. As to the very small organisms, ZoBell thought that bacteria, being small and with every part close to the surface of the cell, would not be much affected by pressure, because the gases could escape through the cell wall. Before embarking on the *Galathea,* ZoBell, who specialises in marine bacteria, tried a series of experiments to prove that bacteria could live under any pressure likely to occur in the oceans, even the deepest ones. He found that, although there were differences in some details of their nutrition between high- and low-pressure cultures, most bacteria could withstand great variations. It was found during the

*Galathea* cruises that small animals such as simple sponges, crustaceans (shrimp-like animals such as copepods), which lived in the depths, presumably fed on bacteria and other microbes. It may reasonably be deduced that there is a whole food-chain from bacteria to fish in the ocean abysses, and the *Galathea* expedition filled in some details.

ZoBell made a study of the varieties and behaviour of the bacteria collected from the abysses. Some of these organisms had different behaviour patterns from those that are to be found in shallow waters or in the upper parts of the ocean. He coined the term 'barophilic' or 'pressure-loving' for these bacteria, and regarded them as being distinct from those living normally near the surface[1]. It seems, however, that these deep-water bacteria are merely normal marine bacteria which have become adapted to living under pressure, and have modified their behaviour to suit their environment. In his studies of the barophilic bacteria, ZoBell used a series of stainless steel pressure bombs filled with a mixture of glycerine and water. He put his cultures in small tubes with neoprene stoppers, making sure that there were no air bubbles to alter the pressure. He then used a pump made from a hydraulic jack such as one finds in motor garages to raise the pressure. One would expect a big pop when the pressure is released, but, because of the small compressibility of water mentioned earlier, there is merely a disappointing fizz. From Dr. ZoBell's work, we know something about the behaviour of bacteria under pressure, but very little about the protozoa and plants; this technique is not adequate for them, because of the difficulty in maintaining sterility when studying plants, and because conditions favour the growth of bacteria when present.

ZoBell and his students showed that pressure altered many of the biochemical reactions within the bacterial cell, and particularly the rates at which they occurred. The enzymes too seemed to behave differently in many cases. The study of pressure effects on living organisms in the seas is difficult because increased pressure alters the concentration of different gases that will dissolve in the water, particularly oxygen and carbon dioxide, both being very important in nutrition. The rate of photosynthesis is dependent on the amount of carbon dioxide dissolved in the water and that of respiration on the amount of dissolved oxygen, so if we alter these by changes of pressure, we will also alter such things as the compensation point. We also alter chemical equilibria because chemical reactions tend to produce substances with less volume as the pressure is raised. This is only logical, since the substances with a small volume will take up less room physically. This fact is used in commercial chemical industry, some reactions being carried out under pressure, others in a vacuum, particularly distillations. Pressure too will alter the viscosity of liquids, so that protoplasm will move more sluggishly and substances will pass through it more slowly.

There are thus a number of factors that are altered, especially in living matter, that make a study of microbes under pressure difficult and unpredictable. The only methods available are those of trial and error, and we really need to devise techniques that can be carried out in the oceans at the depths which we are investigating. With modern techniques, we can in fact do this. We can take samples at any depths we choose, add chemicals or organisms at these depths with special equipment, stop the reactions when we wish, and bring the samples back to the surface for examination. This kind of work is expensive, and that is why it has not been done more extensively. ZoBell and I brought the samples to the surface, which may well have caused changes in the behaviour of our microbes and killed a number of them, perhaps of the most typical barophils. We then re-pressurized them, a procedure which also may have had bad effects. Probably, we did get some organisms which were not unduly affected by this rough treatment, and the findings are at least a good approximation to reality; but they have been questioned, and we do not have the complete answers to such criticisms.

**Temperature**

Temperature is very important in all chemical reactions. Usually the rate of collision between molecules and therefore of reactions, is increased because of the greater speed of molecules as temperatures are raised. There are a few exceptions to this where other factors overcompensate. The solubility of common salt (sodium chloride) in water, for example, is constant with increase or decrease of temperature. Another factor of prime importance in the water environment is the expansion of substances when heated; hence the use of copper rods in thermostats. That water expands with heat is shown by the student's experiment where he heats a flask of water with a long tube through the cork. The water expands up the tube as heat is applied. This happens also in the sea, and as a result the same volume of water will weigh less if the temperature is increased. Thus, increase in temperature results in a decrease of density, i.e. the same volume of water becomes lighter.

This is very important in the oceans and is one of the causes of ocean currents. Surface waters tend to be heated by the sun and thus to become less dense and to stay on top of the cooler water. On the other hand, sunlight also causes evaporation from the surface, and this will increase density while lowering temperature, the principle used commercially in refrigerators and air-conditioning units. Increases of temperature also decrease the viscosity of liquids, and this is important to our microbes as it affects the cell sap and protoplasm, the rate of metabolism and the rate at which materials can pass in or out of the cells through the cell walls. Another significant property is the change in the behaviour of water with changing temperatures, the ones that mainly concern us being 15°, 30° and 45°C. Microbiologists consider that 15°C divides the 'cold-loving' or 'psychrophilic' microbes from the 'mesophilic' ones which live between 15 and 30°C, while the 'thermophilic' or 'heat-loving' types are happiest above 45°C. These somewhat arbitrary figures were set before chemical physicists had studied the changes in the water itself. However, the reaction of microbes to changes of temperature is far from uniform. Some are very sensitive, while others can accept large changes, though time is an important factor. One can raise temperatures gradually with little effect in cases where the same sudden change would be fatal. Many marine microbes, particularly the bacteria, will die at the human body temperature of 37°C (98°F), but they can tolerate low temperatures fairly well. The temperature of oceanic waters varies from over 30°C in the tropical surface waters to above -4°C in antarctic bottom waters, which do not freeze at 0°C because of hydrostatic pressure and high salinity. Most reaction rates are roughly proportional to temperature, and microbes living in colder waters are often sluggish in their reactions though sometimes they combat this tendency by changing their metabolism. This is particularly the case with the 'psychrophils', which, in some aspects at least, change their enzymes below 15°C and thus maintain a reasonably rapid metabolism.

The effects of temperature on biological reactions are affected both by pressure and, in the sea, by salt concentration. Laboratory studies on the effects of temperature alone are therefore not valid. When we include pressure, salinity and time, interpretations of a complicated experimental system become difficult.

Some representatives of the more primitive microbes such as the sulphur bacteria and the blue-green algae have a predilection for hot conditions and live in hot spring areas such as the Wairaki Valley in New Zealand, and Yellowstone in U.S.A.

**Salinity of the Oceans**

Because the main salt in the sea is sodium chloride, it is usually considered for practical purposes that the amount of chlorine in the water is a measure of salinity. Strictly speaking it is called 'chlorinity' and used to be measured by a chemical titration with silver nitrate to give a precipitate of silver chloride. Nowadays, the salinity is measured by the electrical conductivity of the water, and this gives a much better indication of the total inorganic salts present, because it is the components of these salts (known as ions) that conduct an electric current. Nearly all the known elements are to be found in seawater, many of them, such as gold and platinum, in very small amounts. The most important from our point of view is sodium chloride, which varies from about 39 per cent in the waters of the eastern Mediterranean to about 32 per cent close to the ice edge in arctic and antarctic seas and may, of course, be lower in estuaries and river mouths. The amount of dissolved salts in seawater is governed by several factors, the main ones being evaporation and run-off from the land. For this reason the waters of the Arafura, Timor and Sunda Seas have a low salinity due to run-off from an area of heavy rainfall, while the eastern Mediterranean has a high salinity due to intense evaporation coupled with a low run-off. In parts of the western Atlantic there are lenses of low-salinity surface water due to heavy rainfall directly onto the sea itself. Salinity is also increased in low-temperature waters since the ice derived from seawater is fresh, and the freezing leaves the salts behind in the water. However, when seawater is blown over pack ice as spray the salts remain in the ice, forming pockets.

The other inorganic constituents of seawater are derived from the land or from volcanic eruptions. The most important of these include sulphates originally from volcanic activity, calcium (largely from organic materials such as corals, foraminifera, shellfish), magnesium, potassium, nitrate, phosphate (usually controlled by biological activity), iron, manganese, and other elements. The calcium and boron provide chemical equilibria which act to buffer the seawater at a pH of 8.0 to 8.3, and this is extremely important to life in the sea. The activity of boron is almost independent of biological activity, but the calcium carbonate-bicarbonate buffer is of the greatest biological and geological importance.

We are considering salinity with the physical properties because two important properties of salt in the sea are due to its physical behaviour. The first is the effect of salinity on the density of the water. We have seen that increases of temperature decrease the density, and it should be fairly obvious that increased salinity will increase the density. We thus have two factors controlling the density, and both of these are highly significant in the movement of ocean waters, especially vertical movements, both up and down, as in 'upwelling' and 'downwelling'.

The second important effect of salinity is due to the influence of salts on the permeability of membranes. If a cellulose bag is filled with fresh water and put in a bowl of highly saline water, after a time the water on both sides of the bag will have the same salinity. The pressure of the salt molecules on the outside of the bag is greater than the pressure inside the bag, so the molecules go through the bag. This is known as 'osmotic pressure'. If the same experiment is done with a cell such as the seaweed *Valonia,* the higher salt content outside will result in pushing the cell membrane away from the cell wall and the result is what is known as 'plasmolysis', and may even kill the cell. The cells of *Valonia* may be an inch or more in diameter, so this can be observed by the naked eye. Osmotic pressure, like hydrostatic pressure, is a physical phenomenon, though its prime importance is biological. Sharks and sting-rays, which have no swim-bladder, have a high content of urea in their flesh and blood – up to 2.5 per cent – and this serves to increase the osmotic pressure inside their bodies to counteract the hydrostatic pressure outside as they dive to deeper waters. It is because of these 'osmotic' effects that many marine organisms cannot live in fresh water and vice versa. The green algae *Spirogyra* cannot live in seawater at all, and this applies to many of the green 'seaweeds'. The brown and red seaweeds cannot live in fresh water and are confined to the marine habitat. Some species, however, are able to live in either fresh or salt water and some can live in saturated brine, e.g. the purple flagellate *Dunaliella salina,* which gives the reddish-purple colour that one sees on salt lakes. Such organisms are called 'halophils' and usually have difficulty living in seawater because it is not salt enough. There are animals such as the salmon and some of the mullets which move at different stages in their lives from salt to fresh or fresh to salt water. The salmon goes up fresh rivers to spawn, while the mullets usually live in fresh water for a part of their lives, but must reach the sea to spawn, and will die out eventually if they are unable to do so, e.g. if the lagoon that they are living in becomes cut off from the sea.

**Density and Sigma T Diagrams**

In oceanographic parlance, graphs of combined temperature and salinity data are called Ts diagrams or sigma T. They really express density, the temperature increases causing a lowering of density and salinity increases raising it.[3] This means that the water becomes more or less buoyant for the organisms in it as the Ts curve changes, so that they have either to rise or fall or else move voluntarily to maintain themselves in water of constant density, if this is important to them. We frequently find, both in lakes and in the seas, that there are parts where there is a rapid change in density, usually first observed in temperature changes. The temperature plotted in profile against depth usually decreases from the surface downwards, gradually at first and then rather suddenly, then more gradually again. The depth where the sudden decrease occurs is known as the 'thermocline'. There may be two or more thermoclines, but one is usually much more obvious than the others. Sometimes the surface water has a lower temperature than the water below it and we get a bulge between the top and the thermocline. This is known as a 'temperature inversion'. The salinity usually increases from the surface down and we get a change in the rate of increase corresponding to the thermocline; this is called a 'halocline'. It is obvious that the lighter water should always be on top of the heavier water, and this is usually the case. When there is an exception, the system is unstable, and the heavier water tends to sink through the lighter. The Ts curve expresses density relations to depth and where there are sudden changes in density we have what is known as a 'pycnocline'. There may be many pycnoclines in a profile of the oceans and usually each pycnocline will be found at a more or less constant level in a given region, so that the pycnoclines when shown in a solid diagram would form a surface known as an 'isopycnal surface', i.e. a surface of equal density. It might be thought that, in an easily miscible liquid such as water, these isopycnal surfaces would be merely theoretical. In fact, they have the limiting effect of a mechanical boundary to many things such as particles and microbes, which tend to congregate in their vicinity. Because these surfaces have such a marked effect, we get the concept of 'water masses' or 'water types'. These water masses may be very similar or very different in character, and they are part of the system of ocean currents such as the Gulf Stream with which we are all familiar, in theory at least. You will see that evaporation, run-off from land, rainfall, formation of ice and so on can affect the density of the water, and assist in the formation of isopycnal surfaces and thus of water masses. This is why the actual course of surface water masses is often unpredictable. All these are related to climate, not only over the water, but also over land. In the eastern Mediterranean, where much of the surrounding land is desert, there will be little run-off and much evaporation, but the climate and therefore the water structure will be reasonably constant. Along the east coast of America, however, there will be considerable variation in rainfall, direction and speed of the winds, and differences in the amount of ice coming down from the Arctic in different seasons and in different years. Sometimes there is a number of tropical storms and hurricanes starting in the tropical Atlantic and moving along east of the coast of America as far as Cape God, and in other years there are few such storms, and most or all of them cross Cuba and move into the Gulf of Mexico. All these differences in climate will have a marked effect on the movement and direction of the Gulf Stream, and some of them on the intensity of the Labrador current, both of which are essentially water masses.

### The Chemistry of the Seas

The elements Calcium, Magnesium, Sodium and Potassium are very important to the microbes, and the ratio of these to each other is also important, and probably causes the difference between freshwater and seawater microbes. Some microbes need and concentrate strontium and others copper, while all need vitamin $B_{12}$ (cobalamin) which contains cobalt. Bacteria and some microscopic algae can manufacture this substance, but others have to get it from the seawater or from the organisms that do produce it. It is generally accepted that there is enough of this substance normally in the sea for large growths of phytoplankton. Of the non-metals, carbonate (washing soda or sodium carbonate) and bicarbonate (baking soda is sodium bicarbonate) are required by all plants, and produced in respiration by both animals and plants, as we have seen. Nitrogen, as nitrate or ammonia (rarely as the lower oxide, nitrite), is needed by all plants, though some can substitute organic nitrogen such as we find in simple amino acids, breakdown products of proteins. Nitrogen gas is easily dissolved in seawater and can be used by some algae, mainly blue-greens, and bacteria, particularly the anaerobic ones. Other algae can use ammonia, and still others, particularly the diatoms, prefer nitrates. This is interesting, because nearly all the nitrogen present in plants and animals is in the form of amino acids and their more complex relations, the proteins. In these, nitrogen is linked to hydrogen, so that much less energy is required to synthesize amino acids from ammonia than from the nitrate; the oxygen of the latter has to be replaced by hydrogen in the plant cell. It is generally believed that nitrates are important in limiting the growth of phytoplankton. However, Dr. Walsh found that in the Antarctic, nitrate is rarely if ever limiting; there is nearly always enough to go round. He reached his conclusions by field observations and computer analysis, but has not yet made similar investigations in warmer waters.[4]

Microscopic plants require extremely small amounts of phosphate and many can store phosphate against the time when they will need it. It was formerly thought that all plants needed inorganic phosphate as sodium or potassium salts, since calcium and most other phosphates are insoluble in water, or nearly so. Actually, iron phosphate though regarded as highly insoluble, is sufficiently soluble to be used as a source of phosphate for bacteria in culture. It is now accepted that many microbes can use organic compounds of phosphorus and do not necessarily require inorganic phosphate derived from bacterial decomposition of organic material. The blue-green algae *Trichodesmium* lives normally in waters that are very low in phosphate; the amount of phosphate in the water would have to be very low if it were to limit the growth of microscopic algae. Sulphur is present in seawater as sulphate, and plants absorb it as such, but turn most of it into a hydrogen compound known as sulphydrul. This occurs in a habitat containing oxygen, and therefore considerable energy is required.

Some marine plants store substances like bromine and iodine, and one even stores sulphuric acid in considerable strength, but they are mostly macroscopic algae.

## NOTES

1. ZoBell, C.E., and Morita, R. Y. 1956. 'Barophilic Bacteria in Deep Sea Sediments.' *J. Bacteriol.* *73*: 563-8.
2. ZoBell, C.E., and Johnson, F.E. 1948. 'The Influence of Hydrostatic Pressure on the Growth and Variability of Terrestrial and Marine Bacteria.' *J. Bacteriol.* *57*: 179-89.
3. Sverdrup, H. U., Johnson, M. W., and Fleming, R. H., 1942. *The Oceans.* Prentice-Hall Inc., New York.
4. Walsh, John. 1969. Dissertation, Univ. of Miami.

# 8. OCEAN CURRENTS

Most atlases show the surface currents of the oceans (Figs. 12, 13), and these have been known in some form at least from the days of Eric the Red, and according to Thor Heyerdahl, probably from the days of the Egyptian papyrus boats and the South American balsa rafts.

Currents in the oceans are related to the rotation of the earth, to winds, to the configuration of the land and to the shape of the ocean bottom. The currents are four-dimensional, having length, breadth, depth, and being dependent on a time factor which may, in the case of bottom currents, be as much as a hundred years. The ocean water is stratified by reason of various happenings in certain areas. In the Antarctic, during the winter, the freezing of the surface water leaves the salts behind, thus increasing the density of the unfrozen water. Because of the low temperature, this is the heaviest water in the oceans, and it therefore sinks to the bottom, carrying down with it a lot of living matter, including microbes.[1] The centrifugal motion of the earth's rotation forces this water outwards from the polar regions towards the equator, so that a water mass known as the Antarctic Bottom Water moves north along the bottom. It has a relatively high salinity and a very low temperature. Another water masss, formed in much the same way, but north of the Antarctic bottom Water and with a slightly lower density, has much the same circulation pattern and is known as the Subantarctic Intermediate Water. This also moves northward, but above and distinct from the Antarctic Bottom Water. There is a similar situation in the northern hemosphere, but because it is largely land-locked by the Eurasian and American continental masses, the Arctic bottom water is much weaker, and the antarctic and subantarctic water masses actually move into the northern hemisphere. Antarctic Bottom Water is found in the Planet Trench in the Solomon Islands just south of the equator, and in the western Atlantic Ocean east of the Leeward Islands, north of 15°N. The Subantarctic Intermediate water even penetrates into the Caribbean Sea. The Antarctic Bottom Water and the Subantarctic Intermediate Water are, of course, produced by 'down-welling' of surface waters. The Antarctic Surface Water, which has a moderate salinity and a moderately low temperature, moves partly around the world below Africa and South America in what is called the Circumpolar Current, and partly northward along the west coasts of the continents, being scooped up by the intervening land masses.[2] This movement is produced by the rotation of the earth. There is a similar

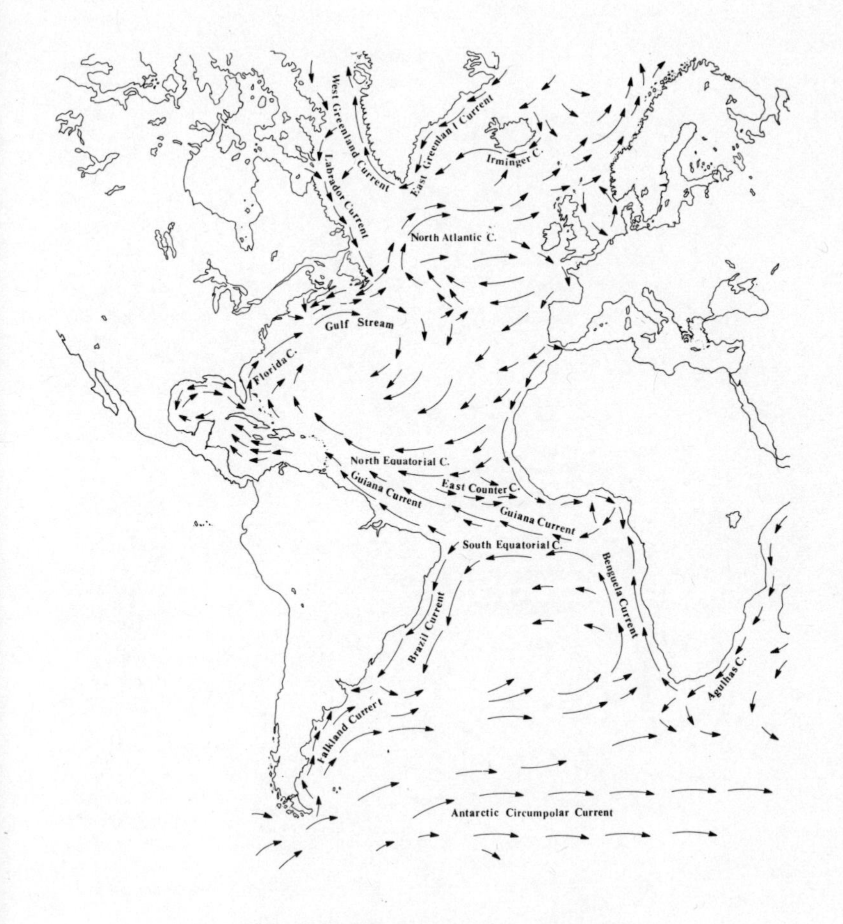


Fig. 12. Surface currents in the Atlantic

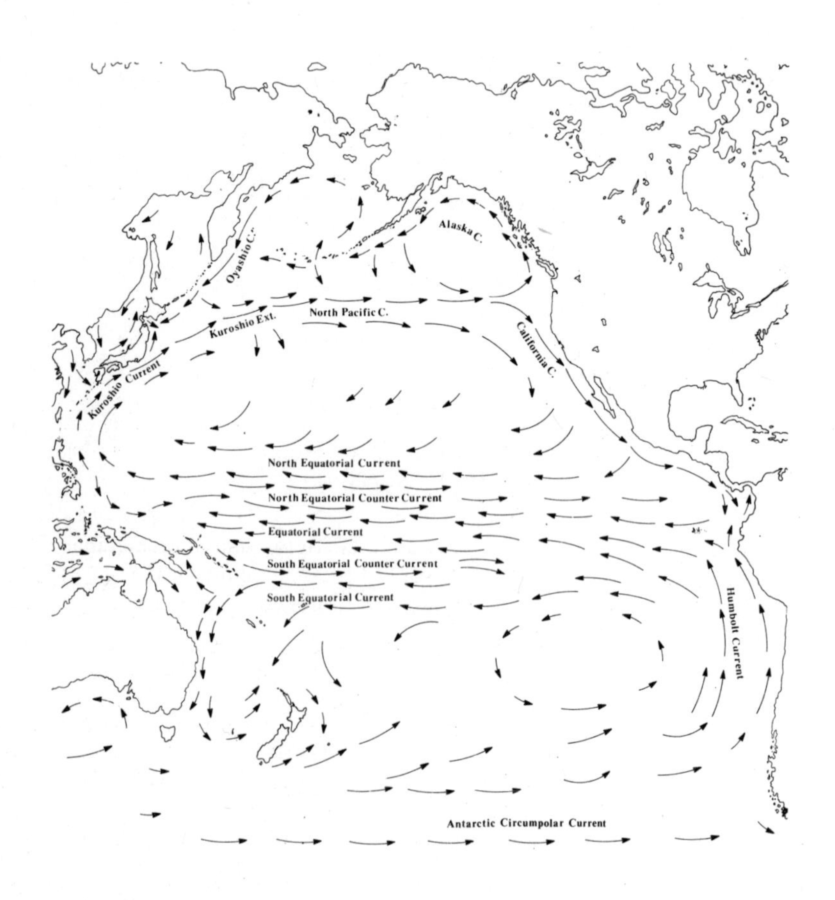


Fig. 13. Surface currents in the Pacific

phenomenon in the northern hemisphere, but no surface circumpolar current, because the Northwest Passage is icelocked. Vestiges exist in the Labrador Current moving along the coast of Labrador southeast to meet the Gulf Stream, and the Oro Shio Current moving south along the coast of Hokkaido and northern Honshu. These currents move south and then east, returning north and westward, and have no counterparts in the southern hemisphere.

The tropical and subtropical surface waters are largely under the control of the trade winds which blow from east to west and move the water in that direction. Thus, we tend to have water masses moving from east to west north and south of the equator and being turned south or north by the land masses. These are the North and South Equatorial Currents, and form generally circular current systems such as the Gulf Stream or the East Australian Current system.

The effect of land drainage is seen in the south-western Pacific where the heavy rainfall in the New Guinea-East Indian Archipelago area produces a low-salinity surface water. This water moves in season through Torres Strait and south from New Guinea, New Britain, New Ireland and the Solomon Islands into the Coral Sea, which in the northern part consists of low salinity surface water. East of this, where there are no large land masses to catch the rainwater and where insolation is high, evaporation produces a much higher salinity water which moves into the Coral Sea and northern Tasman Sea south of the main Coral Sea water mass. These mix to a certain extent off the east coast of Australia, but there is usually a rather streaky and unpredictable pattern of higher (35 to 36%) and lower (33 to 34%) salinity water masses giving a very complex pattern overall to the East Australian Current.

Sandwiched between the surface waters and the cold, bottom waters are a series of water massses, the number depending on more localized conditions. In enclosed areas such as the Banda and Arafura Seas, there may be as many as six narrow bands of water flowing in more or less different directions, one over the other. This makes for an extremely complex system. I have mentioned the presence of two surface water masses in the Coral Sea and partial mixing of these due to a difference in density. Where complete mixing occurs, we call this a 'turnover' or 'overturn', but we may also get a downwelling so that the heavier water mass sinks below the lighter one as happens in the western tropical Atlantic, as we shall find later when we discuss the Caribbean Sea. Sometimes there are even games of leapfrog played by ocean currents in certain areas such as the eastern Bass Strait between Australia and Tasmania.

We have mentioned upwelling (Fig.14) and downwelling, and ascribed these to differences in density between water masses. We have also

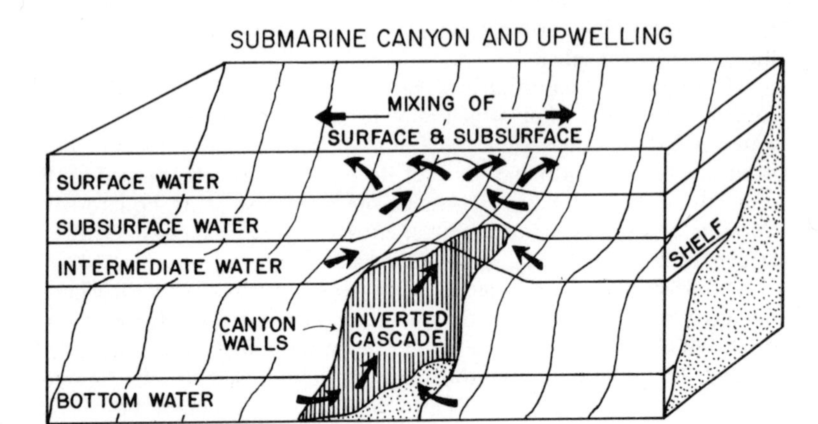


Fig. 14. Upwelling in a submarine canyon

discussed upwelling in local areas as caused by the presence of submarine canyons. In the latter case, portions of a lower and denser water mass are forced upwards by the topography of the bottom and ultimately mix with the upper layers of water. In the former case, it is usually the whole water mass which rises or falls to move over another water mass and this forms what may be termed a 'front'. The dividing line or area between such water masses is called a 'convergence' or a 'divergence' according to whether the surface water is moving towards or away from the front. A divergence is caused by upwelling since the upwelled water must go somewhere and the only way for it to go is outward. For the same reasons a convergence is due to downwelling.

The movement of water masses in an area such as the western tropical Atlantic and the Caribbean Sea will show us the effect of surface and bottom currents on the area as a whole. In the Caribbean, the configuration of the ocean bottom has an overriding effect on the waters entering, and, as a result, on the microbial activity therein. The Antarctic Bottom Water and the Subantarctic Intermediate Water entering the northern hemisphere in the Atlantic and Pacific Oceans, move westwards below the continental shelf of Brazil, but, because the bottom shallows as we approach the West Indian islands chain, the Antarctic Bottom Water cannot enter the Caribbean. It may well have done so in earlier days, since Dr. Harding Owre of the University of

Miami found in the Caribbean Bottom Water a small shrimp-like animal which has so far been found only in the Antarctic, and the Caribbean Bottom Water has characteristics very similar to those of the Antarctic Bottom Water. However, the Subantarctic Intermediate Water can enter the Caribbean through several passages which are deep enough to allow it to pass without any appreciable upwelling; between most of the islands it is forced upwards by a series of submarine canyons, and pushes the overlying water masses towards the surface, a classical example of submarine canyons initiating upwelling.

In this area there are also several intermediate layers of water, one of which has a comparatively high salinity and is believed to originate from the Mediterranean through the Straits of Gibraltar, and to be carried westward north of the equator into the Caribbean. Nearer the surface, two water masses are involved. Coming from the Gulf of Guinea and the African coast, and driven westward by the trade winds, is water known as the South Tropical Atlantic Water, and from north of the equator is a parallel water mass known as the North Tropical Atlantic Water. These are very similar in character, the former being slightly denser due to a very slightly higher salinity. Cape San Roque, at the eastern tip of Brazil, acts as a wedge to divide this water mass, sending part of it south towards Rio de Janeiro, and the rest along the north coast of Brazil towards Venezuela and the Caribbean Sea. Right on the equator, this water meets the fresh water flowing from the Amazon; it is pushed seaward to the north, at the same time pushing the Amazon water westward along the coast of Surinam and Venezuela. In this way the north and south equatorial waters meet, and the northern mass, being lighter, moves over the southern one. They are very similar in character, and contain almost the same community of microbes. So they enter the Caribbean together, but, owing to the upwelling along the West Indies, are often pushed horizontally north and south, with the lower strata coming to the surface in between. This is the sort of complex system that the marine biologist is forced to consider if he is to get at an explanation of what is occurring and why. The surface water moves through the Caribbean towards the Strait of Yucatan with several offshoots and swirls or 'gyres' as they are called, and this shallow strait causes further upwelling. The main water mass now moves eastward and some is deflected north along the west coast of Florida into the eastern part of the Gulf of Mexico towards the mouth of the Mississippi, but most goes through between Florida and Cuba and into the Straits of Florida. Here it becomes the headwaters of the Gulf Stream, together with a little water from what is known as the Old Bahama Channel (between the Bahamas and Puerto Rico) and the Antilles Current which flows northward east of the Bahamas.

The water masses affect the microbes, especially the primary producers, in a number of ways. The surface waters actually carry most of the species along with them, and, as long as the waters do not change their properties appreciably, the plant community stays much the same. The boundaries between water masses are particularly difficult for the microscopic algae to cross. In the East Australian Current, for example, the low-salinity water from the New Guinea area and the Arafura Sea has species which are not found in the higher salinity water coming through between the Solomons and New Caledonia or between New Caledonia and New Zealand. Thus, even if lenses of Arafura Sea water get cut off from the main stream as it moves southward, they can be detected not only by their lower salinity but also by the microbial communities that they contain. Likewise, the fate of the Amazon water can be determined not only by the physical changes, but also by the changes in the plant community as this water mixes with the equatorial waters that I have mentioned, and becomes diluted by them. Surface water masses moving east or west do not change their characters very much, but when they move north or south there are considerable changes of temperature, though these are gradual. This means that the microbes that are sensitive to temperature changes will die out, and the number of species in the community will decrease. This occurs in the East Australian Current and in the current that flows south along the coast of West Australia and also in the Gulf Stream after it leaves the coast of Florida. There are, however, sufficient species left to identify the community with the water mass. This has led to the concept of 'indicator species'. I have mentioned that some species occur in the Arafura Sea but not in the southwest Pacific except in association with water derived from the Arafura Sea. Because the indicator species are limited rather strictly by their habitat they tend also to be somewhat sparse in unfavourable situations and often are rather rare anyway. It is therefore preferable to study the communities or groups of species, since the structure of microbial communities in the upper waters of the ocean gives a lot of information, to confirm chemical and physical findings. Nowadays, one can use cybernetics to interpret such community studies.

The greatest use of plant communities, therefore, in the study of water movements and marine phytogeography is in the upper waters of the photic zone. However, even in the deeper waters, plant communities, diminishing in numbers and species, can exist for many years. Dr. John Walsh was able to follow a water mass in the Bellingshausen Sea from the surface to 1,200 metres by the phytoplankton organisms that he found in. I have found antarctic species occurring in upwellings off the coast of New South Wales, after approximately 90 years without light. Individuals could not live so long and must have been reproducing in the depths. Their appearance at the surface is due to the topography of the ocean bottom. Between southern Tasmania and the south-west of New Zealand, the ocean bottom is level at about 15,000 feet, give or take 100 feet or so. Further north, a series of ridges develops between New Zealand and Australia, the number increasing as one approaches the Coral Sea. This means that the volume of water is compressed into a narrower channel, and also the bottom rises to about 9,000 feet. Off the coast of New South Wales there exists a series of underwater canyons like the Grand Canyon, and the whole effect is to force Antarctic Bottom Water on to the continental shelf. It is this sort of upwelling that causes the well-known rough area east of Bass Strait known to sailors as 'the Paddock', especially to contestants in the Sydney-Hobart yacht race. When this upwelling is strong, it causes a strange phenomenon. The upwelled water is on top of the East Australian Current surface water, so the phytoplankton seems to dip down off Cape Howe and return to the surface off the coast of Tasmania. It also seems to cause surface water from Bass Strait to move north at times along the coast of New South Wales as far as Sydney; when this happens, the East Australian Current is pushed offshore, carrying with it phytoplankton from Bass Strait which is very distinct from that of the East Australian Current.

## NOTES


1. Hart, T. J. 1934, 'Phytoplankton.' 'Discovery' Repts. *8*: 188-9.
2. Deacon, G. E. R. 1933. 'A general account of the hydrology of the south ocean. 'Discovery' Rept. 7: 171-238.

# 9. FOOD AND FOOD WEBS

In each water mass a different amount of organic matter and inorganic salts is dissolved; some waters will be well fertilized with plant foods, others more poorly or even barren. There exist desert areas in the oceans as well as on the land. The Sargasso Sea is supposed to be one such desert, and there are others east of Sydney in the Tasman Sea, and west of Western Australia in the Indian Ocean. As the plants and animals grow and multiply, the nutrient values of the water change and so a water originally rich may become poor and infertile. This is the cause of those deserts which are close to fertile areas such as the Gulf Stream, the Coral Sea and the Banda Sea respectively. Large phytoplankton blooms are often associated with land masses in the warmer waters where heavy rainfall assures a steady supply of nutrients derived from the land. Usually, the diatoms are far more abundant in such rich areas as the Banda, Timor, Arafura Seas, the water above the Carioca Trench off Venezuela and the northern Coral Sea; while dinoflagellates take over in waters poorer in inorganic nutrients. An exception is the eastern Gulf of Mexico, off the west coast of Florida, where dinoflagellate red tides are common.

Phytoplankton is never uniform in any given area but occurs in blooms. In cooler waters there are two well-marked blooms every year, a large one in the spring and a smaller one in the autumn, though they vary in their intensity and in the actual time of their occurrence from year to year. Dr. W. E. Allen, who worked at Scripps Institution of Oceanography, said in one of his papers that during twenty-five years of observations, he did not find any day, week or month in which a bloom could be found regularly enough for a prediction. In the tropics and subtropics there may be a spring or fall bloom, but more usually there are a series of lesser blooms, often occurring with a rhythm which seems to be tied in with solar and lunar periodicity in a rather complex pattern. Sometimes a bloom may consist of one or two species dominating or partly dominating the rest; at other times, many species may occur in a bloom. On one occasion, I counted over sixty species forming a bloom with no recognizable species. There is usually a succession of species each dominating in turn. It has been found that some species excrete substances that are antibiotic, preventing the growth of other species, and finally produce substances that limit their own growth. On other occasions, one species will exhaust one nutrient or produce a substance that encourages the growth of its successor.

Dr. Evelyn Hutchinson, the well-known ecologist, showed in one case that the diatom *Fragillaria* grew abundantly until it had used up all the available nitrate, and then it died down, leaving the field to a blue-green algae, *Anabaena,* which could use ammonia and could also fix nitrogen, but dislikes nitrate.[1]

Usually the inorganic salts required by the phytoplankton are measured in terms of available phosphate and nitrogen just as they are in soils. However, fertility may be controlled by other factors as in the Coorong of South Australia, where, as stated elsewhere, the trace elements copper and cobalt were almost completely lacking and their addition changed the whole complextion of the area.

All the algae excrete dissolved organic substances, which can be used by bacteria and to an extent by other plants. Some phytoplankton microbes excrete up to 50% of their total intake, and the average has been estimated at about 25%.[2] This seems to be a wasteful way of life, and it is difficult to see a reason for it.

The plant material is eaten by animals including certain fishes such as the anchovies and sardinellas of the oceans and the shads and mullets of the estuaries. Most of the plant material however is consumed by small animals which form what is known as the zooplankton, and these in turn are used as food by larger animals, up to and including whales. The most prolific plant-feeding zooplankton are the copepods, and many studies have been made of these, particularly one known as *Calanus finmarchicus* which is abundant in the North Atlantic. There are often large numbers of protozoa associated with phytoplankton blooms, and some of these have symbiotic protophyta lodging within them, thus contributing to the primary productivity. No estimate has been made of the amount of phytoplankton that is consumed by the protozoa either when it is alive or after it has died. From my own observation, in the western Atlantic a phytoplankton bloom is often followed closely by swarms of tiny colourless flagellates that can just be seen at a magnification of 1,000 under the microscope. Ciliates such as the tintinnids and many amoebae also accompany phytoplankton blooms and probably consume a large part of the crop. Numbers of them are found in the stomachs of zooplankton such as the salps, rounded lumps of jelly that at times make the ocean look like tapioca pudding. The salps are interesting animals because they start life as Cordates (with a backbone) and only degenerate later in life into 'spineless jellyfish'. Some of the largest whales and the manta rays feed on zooplankton. Most larval fishes seem to start life feeding on zooplankton rather than on the plants, even if, like the mullets, they become plant-feeders later in life. As fish larvae are very fussy about the size of their food and the smaller larvae need very small food, their preference for a carnivorous diet is strange. In the waters of Peru there live two groups of anchovies, both of the same species; one group feeds on animals and the other on plants.[3]

It is in the photic zone that most of the feeding takes place. However, some of the zooplankton remains always in this region, while others migrate up and down. In many parts of the ocean there is what is known as the 'deep scattering layer', discovered by echo-sounders during World War II as a false bottom in the ocean. At times, it produced quite a strong echo, and submarines used to hide beneath it. It is formed by large numbers of animals which congregate at certain depths. At dusk, a part of the deep scattering layer breaks away and moves towards the surface, but it stops at about 200 feet and stays there all night, returning in the early morning. The place where it stops is usually just where there are most phytoplankton cells, so it is obviously coming up to feed. Actually, the number of phytoplankton cells drops off early in the evening just after the zooplankton comes up, and remains low until after sunrise next day when a rapid reproduction occurs. Dr. Angel of the National Institute of Oceanography in Britain found that one species of plant-feeding zooplankton never comes up into the photic zone, yet has phytoplankton in its gut.[4] This seems to show that phytoplankton which lives in these deeper waters is significant in the food web. However, these phytoplankton may have been derived from the faecal pallets of other species. It is usually conceded that life in deep waters goes on at a very slow tempo, and that all the organisms there are sluggish, because of the high pressure and low temperature.

The efficiency at each level of feeding was believed to be about 10% although Dr. Johannes of the University of Georgia believes that the real efficiency is greater. According to him, some of the bacteria which are ingested with the rest of the food feed on the partly digested material in the animal gut, and after excretion in the faeces continue to break down the material, which is then eaten by other animals, and so on, so that the original food is gradually extracted of most of its nutrients.[5]

When all the plants and animals die, their bodies are broken down, partly by their own enzymes and partly by the enzymes of bacteria, fungi, and protozoa; the residues are consumed by animals of all sizes. So there is a continuous build-up and break-down of organic matter in the seas, forming a continuous cycle. The sudden increase of released nutrients in a water mass usually gives rise to a burst of phytoplankton, which is consumed by the various animals, and it is in areas where this happens that we are likely to find good fishing. Actually, because of currents and the time required for these things to happen, the fishery may be some distance in space and time from the original outburst of phytoplankton which caused it. Increases in nutrients are usually associated with the mixing of water masses, either through upwelling or the impinging of two water masses against each other. Examples of the latter are the mixing of cold water from the Labrador Current with warm water from the Gulf Stream off Newfoundland, of cold, boreal water from the Arctic with the Gulf Stream off Iceland, and the cold Oro Shio and warm Kuro Shio currents off Japan. In these areas, there is very rich fishing in the vicinity of the mixing. In Japanese waters, the relation is particularly striking; as the mixing area moves north to Hokkaido in the summer, and south to the Chiba coast of Honshu in the winter, the striped tuna (bonito) fishery strictly follows the same pattern. The Japanese have made a detailed study of the phytoplankton patterns in this region and have shown striking relationships between phytoplankton communities and the two water masses in all seasons. The Japanese use thermometers to tell them where to fish, and look for sudden changes in surface temperature as indicators of possible fishing grounds; they know from experience that these are boundary layers between currents, and that the phytoplankton and other marine microbes concentrated there lead to good fishing. The most active mixing areas are close to these boundary layers under favourable circumstances.

The outstanding fisheries in the southern hemisphere, such as the anchovy fishery off Peru and the South African pilchard fishery, occur in regions of upwelling of nutrient-rich deep water. The subtropical convergence – where the tropical waters of the Pacific meet the subantarctic surface waters – is very diffuse, extending over 300 miles in breadth.

Dr. Alister Hardy of Oxford noted that large blooms of phytoplankton and swarms of zooplankton rarely occur together, and put forward what he called the 'mutual exclusion theory', that phytoplankton and zooplankton in large numbers exclude each other from the environment. Dr. H. W. Harvey from Plymouth suggested that this was due to grazing and that large zooplankton populations would naturally eat up the phytoplankton as fast as it was produced, and that is why they could not occur together.[6] Now we know that many plants and animals produce antibiotics and, in the case of red tides, actual toxins, so it seems that both these explanations are probable, interacting as circumstances dictate. In the case of heavy phytoplankton blooms, it has been found that frequently the animals, including zooplankton and fishes, are to be found round the edges of the blooms. Until recently, the methods of collecting and estimating phytoplankton have been aimed only at the plants that contain chlorophyll. Now it has been shown that large numbers of colourless microbes, bacteria and protozoa, usually occur around and following phytoplankton blooms, and it seems more than likely that these and the nanoplankton (page 55) are the actual food of the smaller zooplankton.

Of course, the food and the grazing animals must be at the same place at the same time. We have learned that the phytoplankton depend for their occurrence on light and nutrients, and that water masses do not always follow the same pattern from year to year. We can also imagine that if the currents in which the animals are living are deflected from the rich plant areas, there will be starvation. This is believed by many to have been the cause of the disappearance of the sardine from the west coast of North America. In years gone by, the Monterey area depended on the pilchard (sardine) canning industry. For many years now, the canneries have been rusting away. One year the sardines failed to appear, and despite many years of research by marine scientists in an effort to find them elsewhere, no sardines are fished today at or near the Monterey peninsula. Apparently the small sardines failed to reach their food source in time and died out, so there were no stocks of sardines for the years to come.[7]

In 1946, the sardine fishery of Hokkaido, Japan failed on the east coast around Mururan Bay, but the herring fishery on the west coast was very good. The local Japanese moved their boats from the east to the west coast; they knew the vagaries of the currents, and predicted that this change would last for a year or so and then be reversed. They were right, but in this case the locals knew that the stocks of young fish would not be permanently affected.

When the Nile fails to flood, there is little or no sardine fishery off the mouth of that river, and it is likely that there will be a poor tuna fishery around Cyprus because the tuna feed on the sardines. This happened in 1965 when the new Aswan dam was being filled, with the result that the fishermen around Alexandria rioted and many were killed.

In the same area, the Suez Canal is another example, but one which all the effects have yet to be felt. Formerly, no animals or plants migrated through this waterway because there was a series of very salt lakes in the middle. With ships going to and fro and moving the water backwards and forwards, these lakes have become diluted until fish, zooplankton and phytoplankton are moving through – or were until the blocking of the canal. New fisheries have even occurred in the eastern Mediterranean, dependent on fish derived from the Red Sea.[8] If the Suez Canal is never reopened, it may be an advantage from the ecological point of view. It is also a lesson with regard to the proposed building of a sea-level canal through the isthmus of Panama. With the present canal, marine growth is killed off by the fresh water in the locks, so little danger exists.

## NOTES


1. Hutchinson, G.E. 1967, *A Treatise on Limnology.* John Wiley & Sons Inc. New York.
2. Thomas, J.P. 1971. 'Release of Dissolved Organic Matter from Natural Populations of Marine Phytoplankton.' *Mar. Biol. 11* (4): 311-23.
3. de Mediola, R. 1969. 'The Food of the Peruvian Anchovy.' *J. du Conseil,* Copenhagen, *32* (1).
4. Angel, M. V. 1970. 'Observations on the behaviour of *Conchoecia spinirostris.'* *J. Mar. Biol. Assn.* U.K. *50*: 731-36.
5. Johannes, R. E. 1964. 'Phosphorus Excretion and Body Size in Marine Animals.' *Microzooplankton and Nutrient Regeneration Science 146*: 923-4.
6. Harvey, H. W., Cooper, L. N., Lebour, M. V., and Russell, F. S. 1935. 'Plankton Production and its Control.' *J. Mar. Biol. Assn.* U.K. *20*: 407-42.
7. Ahlstrom, E. H., and Radovitch, J. 1970. *Management of the Pacific Sardine in A Century of Fisheries in North America.* Amer. Fish. Soc.
8. Ben Tuvia A. 1966. 'Red Sea fishes recently found in the Mediterranean.' *Copeia 2*: 254-75.

# PART IV

# THE INSHORE WORLD

# 10. THE LIFE OF ESTUARIES

Let us now consider the relationships between plants and animals in the shallower waters where the land has an important influence on both the environment and the animals and plants that live there, particularly the microscopic ones.

In the oceans, the bottom and the top are usually far apart, so marine biologists have only to consider the water as an environment, with the sediments as a special case. In the shallow waters of the continental shelves, and more especially in the bays and estuaries, the bottom is closer to the top, and has to be taken into consideration.

We know what is meant by a *continental shelf* – that shelf around the continents and larger islands which may be broader or narrower, flat or broken, but forms a ledge on which there is usually a quite substantial growth of marine life. It is usually fished, especially by trawlermen, and may or may not be productive, depending on the situation.

The definition of an estuary is not so easy. Some people define it as a river mouth, which is probably the simplest definition. However, rivers often debouch into large bays which may be wide open to the sea, like the Plate estuary in South America, and Rio de Janeiro, which is really not a river at all. The Amazon, on the other hand, forms a huge delta, and so does the Mississippi, but other rivers along the Gulf of Mexico enter almost stealthily behind a barrier island chain, with only a few outlets, not as a rule opposite the river mouths. Chesapeake Bay likewise scavenges a lot of river mouths, but has very different characters from the bays of the Texas coast. It is generally accepted to call all of these estuaries, and make further definitions as required. Whether the North Sea is an estuary is an interesting question, for there is no doubt that it once was when the Thames was a tributary of the Rhine, and the present fishing banks were dry land.

The two main differences between estuaries as we shall look at them, and the oceans and seas are: 1) the salinity and temperature changes in estuaries are much greater than in the seas and oceans; 2) an estuary is much shallower than the ocean. Both these characteristics add up to the fact that the plants and animals in estuaries are exposed to much greater strains and stresses from the environment than the denizens of the oceans. The ocean, like the womb, is a very constant environment, and a comparatively placid one, so the organisms that live there adapt to the constant habitat. The estuary is often subject to sharp changes of temperature between day and night, and of salinity in drought and storm. Moreover, estuaries are strongly subject to terrestrial influences, as for example the effect of phosphate mining in Florida on the eastern Gulf of Mexico, especially inside the barrier islands.

There are two rather specialised types of environment, which have much in common with the estuarine; the ice floes of the Antarctic and coral reefs. However, in both cases the temperature changes are not so great. In the ice environment, much of the plant life is upside-down, i.e. below the ice but attached to it.[2] In coral reefs much of the plant life is associated with the corals, clams and other animals in a symbiotic way.

As there are so many types of estuary, it is very difficult to classify them. Their characteristics depend largely on their geology and origin, the climate both of the estuary and the land surrounding the river valleys that drain into it, and the shape and size of the estuary itself. Their geology and origin have a great influence on their ecology, especially with regard to their microbial life. The long, narrow estuaries of the Texas coast, extending from the Mexican border to Galveston and west past Tampico, are formed by the sea building up low sandbanks across the mouths of normally slow-flowing rivers, and are very shallow. The limars of the north-west coast of the Black Sea are also of this type. The hinterland is dry, with a low rainfall, and the rivers are small, with very little outflow except after heavy rains, when they are subject to flash floods. This means that the estuaries may change in salinity from strongly saline (up to 5%) to fresh within a short period of time. We therefore find mainly organisms which can stand rapid and large changes of salinity, and this reduces the number of species which may be expected. Actually, most of the microbial processes which one finds in the oceans can be adapted to such sudden changes, and one tends to find in the estuaries highly adaptable strains of microbes, but ones characteristic of this environment. They do not seem to relish the placid ocean and quickly disappear in it.

Along much of the east coast of the United States, we find the coastal lagoon type of estuary, similar in general structure to those along the Texas coast (Fig. 15). The coastal lagoons of Florida, including Florida Bay and Biscayne Bay, differ in their origin from the Texas Bays and those farther north, since they were formed by the building of coral reefs and their partial destruction into marls by microbial action. This means that the sediment is mostly lime, and this has a chemical effect on the biology of the region. The lagoons farther north, from Georgia to the tip of Long Island, are more similar in origin to the Texas Bays, but show great differences due mainly to a well-watered hinterland. There is also, as we go northward, a difference in the maximum temperature of the water.

We have mentioned the delta type of estuary, especially with regard

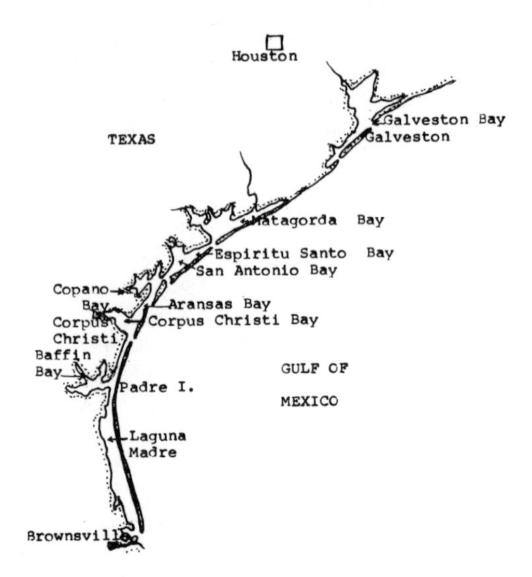


Fig. 15. Barrier lagoons of the Taxas coast

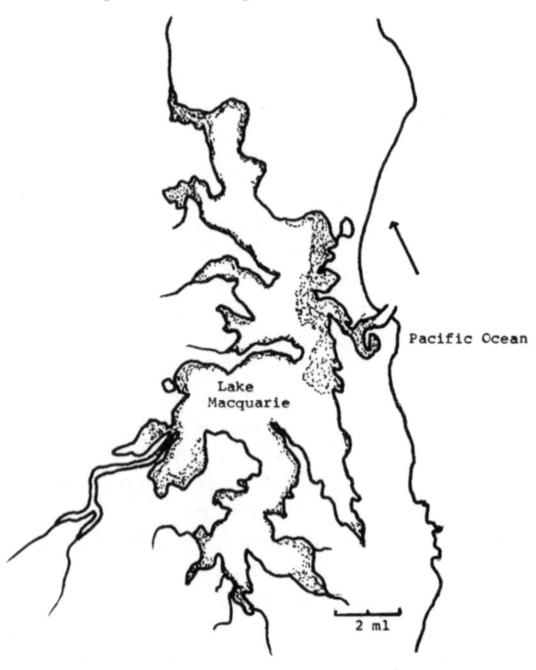


Fig. 16. Lake Macquarie, a drowned valley estuary

to large rivers. These carry fresh water and large amounts of sediment well out to sea, may form canyons through the continental shelf as does the Amazon, or lose their silt by piling it up along the sides of the channel as does the Mississippi, forming what are known as 'silt jetties'.

A common type of estuary is the 'drowned valley' (Fig. 16). This is a valley which has sunk and been flooded by the sea. At times a sand barrier may be formed across its mouth. Drowned valleys come in all shapes, from shallow valleys with gently sloping sides to vertical-sided fjords. In deep estuaries of the fjord type, such as the Norwegian fjords, or Milford Sound in New Zealand, the microbial flora is mainly oceanic in character because these estuaries are deep, and the fresh water which flows from the mountains remains most of the year on the surface. The water below is very constant in its oceanic character, and there is a strong thermocline where the salt and fresh water are in contact. Usually, oceanic water entering the drowned valley will cascade to the bottom, but at times, due to wind or current movements, it will fan out on the surface and with evaporation may become heavier than the water below it.[3] Then we get mixing, or what is known as a 'turnover'. Below the thermocline, when the estuary is stratifield, there is what is known as the 'hypolimnion', the bottom layer, and this normally contains a microbial flora of oceanic origin. When the turnover occurs, this mixes with the surface waters and frequently produces a bloom of phytoplankton. The mixing in an estuary, however, is not as regular as in freshwater lakes, where the turnover is controlled by evaporation and temperature changes only. Between these two rather extreme types, there are all gradations. In the deeper types of drowned valley, especially in the fjords, the water is clear and usually contains very little silt and light penetration is good. In the shallow types, the clarity of the water varies with the wind and the water flow, and the nature of the bottom. In sandy estuaries the water is usually fairly clear, as silica, of which the sand is composed, is heavy and sinks rapidly after it is disturbed. When the bottom is silty, or in limestone areas where it usually consists of fine particles of lime, the water becomes cloudy, even after a slight breeze. This means that the photic zone, or zone of light penetration for active photosynthesis, will be shallower in shallow estuaries and deeper in the deep ones. In the mouths of large rivers such as the Amazon, Congo and Mississippi, the silt load of the rivers makes them opaque almost from the surface, yet there is usually quite an extensive microscopic plant flora in what must be very low light intensity. In the Amazon estuary there is a very peculiar occurrence. The mud consists of two layers, one about ten feet deep, light brown and fairly mobile, the other is dark brown and treacly and squirts out from the ship as she rolls, just as though she was sitting on a muddy bottom. It seems that the treacly layer flows slowly down the Amazon canyon to the north-east, exuding water slowly while the upper layer is pushed to the west by the equatorial currents that we have mentioned and quickly mixes with them. There are microscopic plants in both layers, which is rather unexpected.

## NOTES

1. Lauff, G. (Ed.) 1967-8. *Estuaries.* Amer. Assn. for the Advancement of Science. Vol. 2: 291-302.
   American Fisheries Society. 1966. *A Symposium on Estuarine Fisheries.* Special Pubn No.3, supplement to Trans. Amer. Fish. Soc. *9* (4).
2. Bunt, J. S. 1966. 'Microalgae of the Antarctic pack-ice zone.' Symposium on Antarctic oceanography, Santiago: 198-218.
3. Saelen, O. D. 1967. 'Some features of the hydrography of Norwegian Fjords.' *Estuaries* (Ed. G. Lauff) Amer. Assn. A.V. Sci., Washington D.C.: 63-70.

# 11. COLOURED MICROBES IN THE ESTUARINE ENVIRONMENT

In the average estuary into which a stream debouches, there are definite relationships between the freshwater microbes and those derived from the ocean. The fresh water from the stream flows over the salt water from outside which forms what is known as a 'salt wedge' as it tapers upstream along the bottom (Fig. 17). In floods, the fresh water pushes the salt water out to sea, and steepens the gradient of the salt wedge. When there are strong offshore winds, the salt water moves into the stream

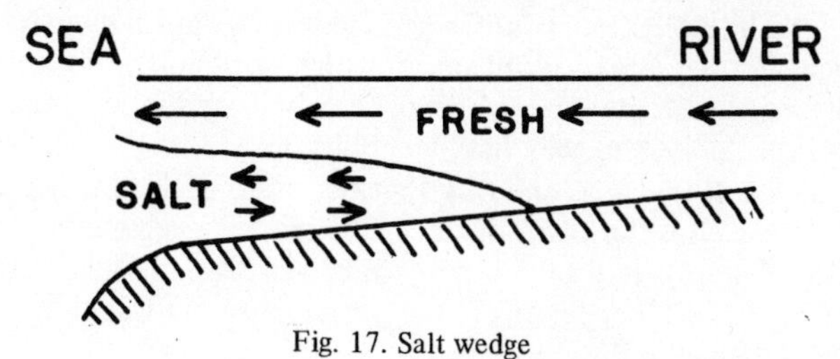


Fig. 17. Salt wedge

and may even push up several miles. During freshets, the freshwater microbes, especially the phytoplankton, move downstream and tend to take over the estuary. With strong onshore water movement, the marine flora moves upstream, especially in the lower waters of the salt wedge, and may for a time establish itself there. Where there is a sill between the sea and the estuary, the ocean water with its flora will cascade into the estuary, and may establish itself in the hypolimnion or lower stratum of water and remain there indefinitely, blooming when conditions are favourable, e.g. when mixing occurs. In lagoons with a small run-off and no strong fresh streams, evaporation from the surface may make the top layer of water heavier than the bottom layer and the water mixes from top to bottom of the lagoon. This also mixes the flora and the nutrients of both layers, and we often get a phytoplankton bloom. In freshwater lakes this occurs in the spring and the autumn and is usually regular; there is also some regularity in estuaries in temperate regions, but very little in the tropics or subtropics as there are so many other factors involved. In highly saline waters, sea water, when it enters, will flow over and not under the lagoon water and we get two marine floras one above the other, the upper one being identical in species with that outside, though possibly enriched. Each estuary has its own pattern of behaviour, and each pattern controls the production of microbes and the kinds of microbes that will appear. The pattern, however, is only a general one, and changes – sometimes drastically – from year to year and even from day to day. In New South Wales, a year occurred when there was twice the usual annual rainfall, and several other years of heavy rains. These followed years when the rainfall was about a third of the average, and the biological response was very different, even in lagoons that were not open to the sea.

Each estuary manages to keep a stock of all the species that are characteristic of it, even though large changes may occur. After a strong freshet, for example, the microbial flora may appear to be an entirely freshwater one, but when the estuary gets back to normal, the freshwater microbes retreat to their former habitat and the marine types appear in their usual place as if by magic. In the drowned valleys the latter can be found in the hypolimnion, while in the shallow lagoons they may linger in the deeper holes or may form spores which settle to the bottom and are available when required. Most things can be found in most places, often unexpectedly, and they have an ability for survival in the face of change. You will find a community that can make itself at home in an environment even if the environment had previously been unfavourable. This does not apply only to microbes; there are tales, some at least authenticated, of fish appearing in closed lagoons after they have been theoretically fished out, or even dried up.

In the southern United States, there are hurricanes, named alphabetically after girls. In 1960, Donna poured salt water all over the Florida Everglades; in 1961, Carla almost destroyed the Galveston-Matagorda area of Texas; in 1964, Cleo hit Key Biscayne, Florida, and Dora Palm Beach; in 1965, Betsy went harmlessly up the coast and then turned savagely on Miami and the Florida Keys. In 1966 it was Inez, and in 1968 the dreadful Camille, which destroyed the Gulf coast of Mississippi; in 1970, Beulah wiped out Corpus Christi, Texas. The hurricanes move seawater over the land and into the estuaries, and the fresh water of the accompanying rains flushes out the coastal lagoons, tearing the environment apart. Donna, however, was the only hurricane that virtually destroyed the mangrove flora of South Florida to the extent that the damage was obvious ten years later. As the hurricanes may hit at any time between June and November, their effect will depend to an extent on the flora which is dominant at that time, and, of course, on the flora which can exist in the time immediately following the disturbance. So we may consider the factors influencing the plankton of an estuary as intrinsic – due to the basic form and history of the area, and to the plants and animals that have developed there – and extrinsic, due to climatic changes, weather, and pollution.

In shallow waters, such as estuaries, there is little buffering of effects such as changes of temperature or salinity, and therefore the time for such changes in very short, hours or perhaps days; whereas in the oceans time is more often measured in months, years or even decades. This means that the microbes must be able to adapt much more rapidly if they are to survive. As they are tiny and have a large surface area compared with their volume, the reaction between the organism and the environment must be quick and complete or the organism will die. These organisms can adapt to changes that would kill you and me, complex and sophisticated though we be.

In estuaries, and on some continental shelves, the microbes of the bottom (benthic) and those attached to larger plants and animals (epontic) are far more important than phytoplankton in the scheme of things. Few fish or other animals, except shellfish such as oysters, mussels and cockles, eat much phytoplankton. They tend to prefer the attached microscopic algae and those on the bottom which are washed about by the wavelets and tides. Tides do not move on to a bank or flat at right angles to the shore, but rather in a vague pattern of backward and forward motions more or less parallel to the shore. The water also moves quite considerably through the sediment between the tide marks, more or less vertically, and this tends to carry many microbes up and down, or at least to facilitate their vertical motion. It is not surprising to find microscopic plants such as diatoms up to 12 centimetres down in the sediment, and quite active microbes at that.[1] The lower ones probably act as a reserve in case the ones in the upper layer are killed off by a sudden frost or some other catastrophe. They do not seem to be affected if there is little or no oxygen, and they can adapt to anaerobic growth (growth without oxygen), another example of their versatility. One diatom in particular, *Hantzschia amphioxys,* can and does migrate through the sediment as the tides rise and fall, appearing on the surface to form a brown scum on the top of the sand when the tide is out, and returning to the sediment as the tide comes in. You can imagine the huge numbers of this microbe required to form a visible coloration on the flats. It starts as a few light yellow-brown patches on the bare sediment and these gradually coalesce to give rise to a most interesting spectacle. It appears after the tide leaves the flat, and strangely enough disappears before the tide returns; there must be a special trigger mechanism or something approaching a time clock. It goes down into the sediment up to a centimetre. In this environment are the bacteria, especially the anaerobes and the sulphur bacteria; and the detritus feeders, the ciliate protozoa, the coloured and colourless flagellates, and other species of diatoms and the fungi. The blue-green algae usually appear when conditions are getting unfavourable for other species except the sulphur bacteria, and bacteria which live without oxygen. One of the commonest places to find blue-green algae in estuaries is in the intertidal area where they form a felt or carpet, greenish or reddish-brown in colour and about

2 millimetres thick. These felts usually split irregularly and curl up at the edges like clay does when it is drying. Frequently there are filaments of these blue-green algae in these drying clays. Another area where the blue-greens are frequent is in sands, also intertidal, especially near the ocean. In the Gulf of Mexico I have seen blue-greens extending at least 2 inches into clear, coarse sand. There are many small animals that can, and probably do, feed on them and that live normally in these sandy shores.

A number of micro-algae normally live in and on the sediments, and only occur in the water when the bottom has been disturbed. If you put some coarse sand under the microscope, using a special combination of filters (BG12 below the condenser, and OG1 on the eyepiece) you will see a larger or smaller number of shiny red dots around the sand grains. These are microscopic algae, mainly diatoms, which are attached to the sand. The red colour is due to chlorophyll, which shows a red coloration under fluorescent light. Being attached to a solid grain has advantages for the algae in that, should the grain be disturbed by the tides or currents, it will sink again to the bottom and the attached algae will be returned to the sediment in which they prefer to live. Along with the microbes that normally live in the sediment, there are others belonging to the phytoplankton and to attached species which have become detached and sunk to the bottom, since they are heavier than water. Many species take advantage of this to endure unfavourable conditions, and quite a few form spores which are usually much heavier than the normal plant. They may stay in the sediments for months at a time, and germinate only when conditions are perfectly favourable. Certain plants, especially desert plants, have the ability to select conditions for the germination of their seeds so as to ensure that they will have time to grow, flower and produce another crop of seeds while conditions are still favourable. This seems, in a simpler form, to be possible for the microscopic algae, and they appear to be able, in some mysterious way, to predict conditions.

Much has been written, and still more conjectured, concerning what are known as biological rhythms. There are, even in humans, strong cyclic changes, frequently lunar. Other cycles seem to depend on solar rhythms, and sometimes there are complicated rhythms partly due to the effects of the sun and partly to the moon. I have even found that the growth of barnacles and other animals on the bottoms of ships and other structures showed such rhythms, and there are always the farmers who plant by the moon. Fishermen also tend to believe in working to the phases of the moon, and such superstitions are usually founded on reliable observation. There are records of rhythms among the micro-algae. Some would appear to have lunar rhythms, especially as regards the germination of their spores. *Hantzschia amphioxys* waits until the tide has left the flats before it appears on the surface, and this could be explained by the drying-out of the sediment acting as a trigger. But *Hanztschia* also disappears before the tide comes in, though of course capillarity may moisten the flat before the tide wets it. Some scientists have studied this phenomenon in the laboratory, and say that the rhythm will persist for a time even if the tidal effect is reversed, but that eventually the organisms will learn the new rhythm and will again take time to relearn the original rhythm.

On the continental shelf we find very similar conditions to those in estuarine sediments, and many of the same organisms occur there, including the micro-algae, but there are fewer species as the water is deeper, and light conditions are not so favourable. On the other hand, changes in temperature, salinity and nutrients are not nearly so drastic. In this environment live the bottom fish which feed on animals and plants associated with the sediments. These fish are often caught in trawls. The trawls stir up the bottom, and, of course, the microbes that are associated with it. In a large, flat area such as the North Sea, these will settle somewhere else, though some will die. Intensive trawl-fishing will cause destruction in one area; but another area where the fishing is less intense may be improved.

Quantitatively, the most important microscopic algae in the estuaries are usually the ones that attach themselves to larger plants and other material. The most important plants for this are the sea-grasses (plants related to the iris), which have returned from the land to make the shallow seas their home. They have flat leaves, so can support a considerable fuzz of tiny plants, often greater in weight than the plants themselves and, as the microbes have a relatively short life, far greater in weight when taken over the whole year. For this reason, the sea-grass flats are the most important fish-producing parts of any estuary. Of course, as the leaves of the grasses die and turn into detritus, the attached plants become part of the bottom flora, or their descendants do, and this adds to the mass of plant material available to the bottom-feeding animals.[2] Among these we include the shrimps and the shellfish – clams and oysters. The Toheroa (New Zealand Clam) feeds almost entirely on a diatom, *Chaetoceros armata,* which gives it its green colour and probably its flavour. Most oysters and clams have a mixed diet of diatoms, flagellates and bacteria which they pump through at a very rapid rate. This causes a suction, so that the oyster and clam can draw some of the algae away from their attachment, as well as sucking up the bottom-dwelling (benthic) species.

The feeding of mullets is very interesting. If you watch them, you will see that they move slowly up and down the leaves of the sea-grasses, sucking away. If you look at the contents of their stomachs, you will find that they have been eating the attached algae, particularly the diatoms. Then they move over the flats and swim around just over the surface of the sediment. In this way, they stir up the bottom, but allow the heavier material to sink again, leaving mainly the plants which they suck in and eat. They also swallow a large amount of fine sand which acts as a grinding paste since these fish have no teeth.[3]

## NOTES

1. Harris, Roger P. 1972. 'Distribution and Ecology of the Interstitial Meiofauna on a Sandy Beach.' *J. Mar. Biol. Assn. U.K. 52*: 1-18.
   MacIntyre, A. D. 1969. 'Ecology of Marine Meiobenthos.' *Biol. Rev. 44*: 245-90.
2. Finchel, T. 1970. 'Studies on the Decomposition of Organic Detritus derived from Turtle-Grass.' *Limn. Oceanogr. 15*: 14-20.
   Darnell, R. M. 1967-8. 'Organic Detritus in relation to the Estuarine Ecosystem.' In *Estuaries.* Amer. Assn. for the Advancement of Science. 1967-8. Vol.2.
3. Hickling, C. F. 1970. 'A Contribution to the Life-History of the English Grey Mullets.' *J. Mar. Biol. Assn.* U.K. *50*: 609-23.

# 12. COLOURLESS MICROBES IN ESTUARIES

The coloured microbes provide energy for the whole system from the sunlight shining down on it. On a sunny day, bubbles arise on top of the mats of algae, whether they be blue-green, green, brown or red. Often, they break away and rise to the surface, disappearing there. On a sea-grass flat, there are even more bubbles, and sometimes they form a steady stream, mainly from the tips of the leaves. These bubbles are oxygen, and the reason they appear is that the water is already supersaturated. During the night, the bubbles cease because there is no photosynthesis, but respiration continues. The pH of the well-buffered seawater in the open ocean is slightly over 8 and is alkaline. In the estuaries it can reach 9.4 on a bright, sunny day on a sea-grass flat where photosynthesis is active; at night it sometimes falls to an acidic state with less than pH 7. The high pH is due to reduction of the carbonic acid content of the water from photosynthesis, and the low pH to reverse changes due to respiration increasing the carbonic acid. This means that the plants, especially the microscopic plants, control the chemical, and therefore the biological processes in an estuary.

In the sediments, however, there is another story. Below their surface there is to be found a wholly different regime of microbes, most of which are colourless, but which are just as important in their environment as the coloured microbes are above them. Most of these colourless microbes, though not confined to the bottom, do not play a very important role in the water.

The chief organisms of the sediments are the bacteria, the fungi, and the protozoa. If you take a culture of bacteria in seawater or in a liquid culture medium, and shake it with fine, sterile particles of detritus or clay, you will find that as the particles settle down, the bacteria go with them, and you may even get a sterile water with all the bacteria taken down into the sediment. Even in the water, nearly all the bacteria are to be found on particles and few are swimming freely. Even the dissolved chemicals tend to become concentrated on any particles present and on the walls of the container. When mud is stirred up, both the bacteria and the nutrients are concentrated in the sediment as it settles down. There is usually a cloud of fine sediment particles in the bottom water when there is any current or tidal action; in shallow waters, any heavy wind serves to stir the sediment. This acts as a natural concentrating agent in shallow waters, and greatly facilitates the activitity of bacteria and other microbes in this environment.

Anyone who has observed the seaside will have noted the layer of dead and dying sea-grasses or dying masses of algae which drift about in the estuaries and finally end on the bottom. If you look at them under the microscope, you will find that they are breaking up, and that there are at first dead and living patches, but finally only dead material, and that they end up as a grey or black sludge, usually mixing in with the sediment. In this mess there are large numbers of bacteria, some fungi, and often a great many protozoa, each actively moving in pursuit of its way of life. The bacteria and some of the fungi will have attached themselves to the dying material very early, and will multiply and increase as death takes over. They digest certain parts of the material, and cause it to break up. When the particles of broken material get small enough, the protozoa come into the picture. If you look at protozoa under the high powers of the microscope, many of them have what looks like a very primitive mouth, called a 'gullet' because it is used to take in food. Usually the flagella or cilia, those tiny organs that thrash about in the water and characterize the protozoa, push small particles towards the gullet, and they tend to be held there mechanically until the organisms can fold itself over them, after which they appear inside and can be digested.

All these processes go on in the ocean, but there the space between the particles is so great and the number of microbes is usually so small that considerable concentration is necessary before they can be studied.

Bacteria have the power to dissolve substances by means of a great variety of enzymes. These enzymes allow the microbes to penetrate the cells of the plant or animal and use many of the organic substances inside the cells. Bacteria can attack and use nearly all these substances, including material like lignin (wood), cellulose (cotton) and chitin (the shell of lobsters etc.), which are very hard to dissolve chemically. It is because of the activity of the bacteria that such substances do not accumulate on the bottom of the seas and estuaries, so the bacteria may be regarded as the scavengers of the sediments. There always seem to be present the kinds of bacteria that can cope with any organic compound that happens to be there, and their numbers are such as to prevent an accumulation of any particular substance. Bacteria are very specialised in the medical context and it has been difficult for bacteriologists, formerly trained in medical laboratories, to accept the concept of omnivorous bacteria. Once such a substance as chitin is available, bacteria can divide every 20 minutes, and very quickly an army of them is available to consume it. Once it is consumed, many die, but some will turn to other foods and remain available for future requirements, or rather their descendants will. Bacteria have heredity in the shape of chromosome material just like the higher plants and animals, and because of their rapid multiplication, the chances of mutations and changes of character are far greater than in the higher plants and animals. As specificity is characteristic of disease-causing bacteria, so versatility is characteristic of marine bacteria, an essential character in an environment that changes so rapidly.

Bacteria may be divided into two groups, those which need oxygen and are called 'aerobes' and those which do not and are known as 'anaerobes' Most of the marine bacteria which can live with oxygen can also live without it, changing their habits to accommodate themselves to the environment. However, many anaerobes can only exist in a resting condition as spores, if oxygen is present. These 'obligate' anaerobes include some rather nasty types such as the tetanus bacillus and the bacillus causing botulism, one of the most deadly poisons known. Both of these have been found in the sea, but fortunately are very rare.

It is the aerobic bacteria which are active in the water and in the upper parts of the sediments. They cause the original breakdown of the plants and animals, using oxygen in respiration as they do so. Although individual bacteria require only very minute amounts of oxygen, there are so many of them, and they are so active when they start the break-down, that the total amount of oxygen used is considerable, and finally it becomes exhausted. Up to this point, limited changes have occurred in the organic matter or detritus, but now the anaerobes are in a position to take over the main role, and the organic matter is reduced to its ultimate constituents. These are gases, such as nitrogen, hydrogen, hydrogen sulphide, methane, and ammonia. Apart from the nitrogen, all the gases contain hydrogen but no oxygen. The nitrogen escapes during the process of reducing oxygen-containing nitrogen compounds such as nitrate to ammonia. Nitrogen, methane and hydrogen sulphide may escape into the atmosphere, but hydrogen and ammonia are too reactive and form compounds in the sediment. Methane is 'marsh gas' and is believed to be the cause of the 'will-o'-the-wisp', those strange lights that appear on the surface of marshes due to the combination of this gas with oxygen – that is, it burns. Hydrogen sulphide can be detected by its smell.

There is an interesting measurement of the presence of oxygen or the lack of it, known as 'oxidation-reduction potential', 'redox' or 'Eh'. There is no adequate meaning or definition of Eh, but if there is no oxygen present and enough hydrogen or hydrogen sulphide, the sediment has a strongly negative electric potential when compared with the calomel electrode (the electrode used for measuirng pH also). If oxygen is abundant in the system, the Eh is positive, and the maximum difference is about 1 volt. So there can be 1 volt difference in energy between the sediment and the surface of the water of an estuary, quite a high figure.[1] It also has a great significance for the microbes and other plants and animals that live in the estuaries. All true anaerobes require a negative Eh of -0.1 volts or more, and it is the function of the aerobes to provide that condition. A number of bacteria in sediments require such conditions and are important. There are the bacteria which convert the sulphates commonly present in seawater into sulphides, first to hydrogen sulphide. Two processes are working here; first, as the Eh is lowered, hydrogen is released from sulphur-containing proteins; secondly, true anaerobes make hydrogen sulphide from sulphates such as gypsum (calcium sulphate). This is the reason for the smell of rotten egg gas in heavy muds, even in areas where there is little or no human habitation. The same smell occurs in the vicinity of sewer outfalls, because of the same processes. However, the smell of hydrogen sulphide does not necessarily indicate that a sewer outfall is close! Usually, when you can smell hydrogen sulphide, it comes from a black mud, sometimes close to the surface, sometimes deeper down under a layer of clear sand. This mud consists of sulphides of iron, which are very insoluble, except in acids. Sometimes this production of sulphide goes on so strongly that even the water in deep holes becomes black and fish die from lack of oxygen or from the poisonous effects of hydrogen sulphide. The same effect occurs in Walvis Bay during the red tides, and is not uncommon in partly land-locked bays; also in freshwater sink holes in the Florida Everglades. An important side-effect is the release of phosphate into the sediment and the water, thus providing on-the-spot fertilization, followed quickly by a large phytoplankton bloom.

Sediments with a black base may have a layer of yellowish green, usually faint, and above that a purple layer that may be quite thick. Sometimes, the purple layer appears on the surface as the tide goes out, in just the same way, and usually in the same areas, as you find the *Hantzschia* on the surface. The green and purple layers are due to the green and purple bacteria, which contain a primitive form of chlorophyll. This bacterial chlorophyll, as it is called, allows the bacteria to change sulphides back to sulphur or sulphates when there is no free oxygen present, for the bacteria are anaerobes. In doing this, they also assimilate carbon dioxide just as the higher plants do. The green sulphur bacteria can only grow and multiply if there is no oxygen, but the purple ones can live with oxygen, but in this case, they act like ordinary bacteria and use neither their chlorophyll nor the sulphides. There is evidence that these bacteria are responsible for sulphur deposits.

There is another interesting process which does not need any living organisms at all. It requires only finely divided iron and sunlight, and this acts in the same way as the chlorophyll of the bacteria. If oxygen is present, as at the surface of sediments, sulphides and sulphur will be converted slowly into sulphate. This is why you can smell sulphur dioxide when the wind is blowing towards you from a stack of sulphur in a fertilizer factory. It is also why you will see a white, milky surface on water when there is a smell of hydrogen sulphide emanating from below. The white material is sulphur. Bacteria known as *Thiobacillus* use this reaction to get energy and assimilate carbon dioxide, and are usually found on the surface of sediments in light-coloured areas.

The purple and green sulphur bacteria, the Thiobacilli, and some strains of the sulphate-reducing bacteria can be grown on very simple culture media which do not contain any organic matter at all – just phosphate, a nitrogen compound, some trace metals and the right sulphur source. Other strains of the sulphate-reducers need a sniff of vinegar (acetate) or lactate (the acid from milk sugar).

Ammonia is formed by bacteria in the sediments. It combines to form ammonium chloride or sulphate and is thus stored. However, all nitrogen compounds are very soluble in water, so they are easily available when they reach the surface; they can be used by some plants as they are, by others when converted by bacteria into nitrite or nitrate.

In these ways, the sediments act as storehouses for nutrients in estuaries, and release them under strictly controlled conditions, that is they control the fertility. So, if we use an estuary for mariculture or for angling, it is very important for us to understand the microbial processes and, if we can, to control them so as to give us the maximum productivity.

The role of fungi in the marine environment is not at all well defined, and it is believed that they are much less important than the bacteria. They require oxygen so cannot grow far below the surface of the sediments. It is thought that perhaps they might be important parasites of animals and plants as they are on land, but this still has to be proved in most cases. The only known case where fungi were suspected of causing mass death of marine organisms was the '*Zostera* wasting disease', which wiped out that sea-grass along the northeast coast of America and parts of Europe in the 1930s. Fearful damage was caused, and as we would expect from what we have said about sea-grass flats, the whole life of the estuaries was brought to a standstill. Recovery was slow – about 25 years – and there is still argument as to whether a fungus was really the cause, or which of two possible organisms, if any. Yeasts are present in seawater in the oceans and the estuaries, and a very unexpected discovery has recently been made about them. Some of the most important are a stage in the life of a smut fungus, similar to those that cause the sticky black dust that you see at times in the seeds of wheat and other grasses. Yet we have no evidence that these fungi are parasitizing animals or plants in the seas.

There are groups of fungi, some of them specially adapted to aquatic life, that are found in estuaries, penetrate the structures of sea-grasses and algae and are responsible for decomposing detritus in the aerobic sediments.[2] Most of the studies of marine fungi have been sponsored by people interested in wooden structures or rope materials not naturally occurring in the seas, except by chance or storm. So our knowledge of fungi in the marine world is biased towards some specialist fungi, and we are still ignorant of the true importance of these microbes in the seas.

You can always find colourless flagellates and ciliates, often mixed with certain forms of coloured flagellates, in the water and in the sediments. Many of the coloured flagellates are very closely related to the colourless ones and their possession of chlorophyll can be temporarily or permanently stopped experimentally in the laboratory. If we look at sediment under the microscope, we will see large numbers of ciliates in particular, including *Paramecium,* madly looking for food, and apparently scraping that food off sand grains and other particles. Flagellates do the same thing, but are slower about it, due to their less efficient feeding mechanism. The end result is the same; they consume a lot of detritus and a great many bacteria.

Some ciliates can live anaerobically, and in such conditions are very numerous and active.

There are often quite a few amoebae in the water and sediments, though they do not seem to swarm as they do in the oceans. They also consume particles of organic matter and engulf them without difficulty.

It is quite obvious that these colourless organisms must collectively play a very important part in the biology of the sediments and the area immediately above. It is difficult to show that they are eaten if we examine animal stomachs because they are easily broken up, and lose their identity. However, they have not yet succeeded in overrunning the entire estuary, so there must be a strict population control placed upon them by nature. We can accept the idea that the burrowing shellfish, clams, shrimps and their allies make great inroads into the populations of the colourless microbes as a steady article of diet. Once again, our knowledge of the habits and role of these microbes is limited by the fact that few people have studied them and nearly all the work has been on classification.

Areas where the bottom consists largely of lime have certain peculiarities. These areas include coral reefs and atolls, and once again their ecology is only scantily known apart from the taxonomy of the animal and larger algal forms. Anyone who has spent time on a coral reef will appreciate that there must be a tremendous life force at work there. Animals, including the attached forms such as the corals, clams and anemones, are all busy feeding constantly, and the fishes, shrimps and other free-swimming forms are always numerous. There are few or no sea-grasses and the larger algae are not very large after all, consisting mostly of red or green coral-like forms collectively known as the 'coralline algae'. Some of the parrot fishes and a few other fish species feed on these, nibbling at them, but they are not available to most of the reef-dwellers.

There remain to be considered the microbes of the reefs. Many of the attached species of animals of the reef are coloured, and microscopic examination of these will show that the tissues contain small, yellow spheres. These are, in fact, microscopic algae which live symbiotically in the animal and which have already been mentioned as 'zooxanthellae'. They form a symbiotic arrangement with the animals, providing, it is believed, sugars and other substances which they make as a result of photosynthesis and exchanging the excess of these substances for things that they cannot make, but which the animals can. In addition, they get protection and are unlikely to be eaten unless the whole animal is eaten. As an additional protection, many of these animals have stings, so are not very palatable. Sometimes the animal will devour its zooxanthellae, taking in a new batch later.[3] At times, some animals such as the pen and fan corals get rid of their plant companions when disturbed. Sometimes they divide up their plant companions before they themselves divide, so that each new animal has an equal share; at others, they eject them and take them back later. The plant cells or zooxanthellae, when they are outside the animal, change their shape, develop a waist and two whip-like appendages or flagella, one trailing and the other rotating round the waist, exactly like the genus *Gymnodinium* of the Dinoflagellates. They are therefore classified with that group.

When we consider the reefs quantitatively, we find a suggestion that the plant material, particularly the microbes, forms about five times as much as the animals.[4] This sort of ratio is necessary, because of the loss at each link in the ecological food chain. The phytoplankton does not seem to be very important, so the zooxanthellae would appear to be the real primary producers of the coral reefs.

The plant and animals of the reef must either be eaten or die and be broken down by bacteria or other microbes, and the little work that has been done showed that a large amount of organic matter of detrital form gets into the crevices of the reefs and is rapidly attacked by bacteria, so much of the destruction occurs on the reef itself.

When carbonic acid gas is passed through lime water, a white precipitate is formed which dissolves again as the gas passes through. This process is continually occurring on the reef and can be due to either the microscopic algae or the bacteria. The blue-green algae are particularly active in it and can dissolve and precipitate lime (calcium carbonate) under differing conditions. These blue-green algae grow abundantly over dead corals, forming what are called 'nigger heads'.

The ice environment of the polar regions has large microbial populations, both on the pack ice and below it, where a felt of about a quarter of an inch may be formed. When the ice melts, these algae will be discharged into the water, enriching it with a large number of individuals and many species to start the spring bloom which is characteristic of those parts. In addition, there is a number of species which are normally found in estuaries, including forms normally attached to surfaces, and this is why I include the ice environment among the estuarine ones.

## NOTES

1. Mortimer, C. H., 1941-2. 'The Exchange of Dissolved Substances between Mud and Water in Lakes.' *J. Ecol.* Cambridge. *28-30.*
2. Wood, E. J. Ferguson. 'Common Marine fouling organisms of Australian waters.' *Austr. J.Sci. 18*: 34-7.
3. Muscatine, L., Poole R., and Cernichiari, E. 1972. 'Some Factors influencing Selective Release of Soluble Organic Matter by Zooxanthellae from Reef Corals.' *Mar. Biol. 13* (4): 298-308.
4. Moore, Hilary B. 1958. *Marine Ecology.* John Wiley & Sons, Inc., New York.

# 13. MICROBES OF THE OLDEN DAYS

## 1. The Origin of Life

The early atmosphere of the earth was composed of gases such as nitrogen and hydrogen, but no oxygen. When the first oxygen was released from rocks, it probably combined with hydrogen to form water. We thus have a picture of the hollows of the earth's surface containing a quantity of water produced by chemical means from oxides, but not life. Other inorganic chemical compounds and gases such as carbon dioxide (formed from carbonates) were released from the bowels of the earth, and in the absence of a protective atmosphere, the ultra-violet light of the sun's rays was able to act as a sort of catalyst to provide the energy for combining some of these carbon compounds into primitive organic compounds. As these compounds became more complex, there were probably formed larger and larger molecules, capable of self-replication and on the threshold of life. The waters on the earth's surface, the primitive ocean, as a result of ultra-violet catalysis must have contained a large number of unorganized chemical entities, forming a sort of organic soup which the Germans call the 'Weltschlamm' or world-slime. Ultra-violet light, which is toxic to living matter, is able to supply the energy for making amino acids from ammonia and carbon dioxide and for piling amino acid molecules together to make more complex substances such as proteins. Somewhere in this mess, ATP (adenosine triphosphate) and its associated compounds were formed. These phosphate complexes are essential to life and reproduction, though it may not be true that if we could synthesize them all and control their production we could create life.

At some stage in the development of the 'Weltschlamm', membranes were formed which separated certain molecules or groups of molecules, and some of these separated groups obtained the power of reproducing themselves in similar patterns. At this stage we had life, a very simple form, but with all the necessary powers to develop into life as we know it.

There is doubt as to what organisms were the precursors of life. Some consider them to have been bacteria, and others plants with chlorophyll. We shall probably never know. It is, however, agreed that the first life must have been anaerobic, capable of existing without free oxygen. Some people think that the bacteria came into existence at this stage and that they lived on the organic soup. I believe myself that they were indeed bacteria, but developed earlier and lived on inorganic transformations such as the oxidizing of hydrogen sulphide, methane or ammonia, to sulphur, carbon or nitrogen. Oxygen would be required to make sulphate, carbon dioxide or nitrate in the first stages of the living world, so such syntheses would not have been possible. Autotrophic bacteria have simpler forms and shapes than the saprophytic anaerobes, which are higher in the scale of bacterial development, and they also form resting spores.

It is easy to follow in theory the development from bacteria to plants and thence to higher forms. We find the link in the sulphur bacteria – Gram-negative non-sporing rods, mostly autotrophs, some with a primitive chlorophyll. From them could come two lines of succession: first the sporing anaerobes, such as *Clostridium nigricans,* similar in shape and appearance to the sporing saprophytes and, like them, Gram-positive; secondly, the thread-like *Beggiatoa* and *Thiothrix* which can use sulphide oxidation or ordinary saprotrophy for their nutrition, and resemble the blue-green seaweed *Oscillatoria.* Likewise, the huge bacterium *Thiovulum* has a similar form to some of the green flagellates. It is significant that the blue-green algae can live in very primitive environments such as hot springs, highly saline waters, and in sediments without oxygen, often associated with transformations of sulphur. In short they can live in situations, possibly resembling those on the young earth, where the only other life forms are bacterial; and like the bacteria, they can tolerate a very wide range of conditions.

Some of the higher bacteria, too, grow like very primitive fungi, and are in fact borderline forms. Thus, from the evolutionary point of view, there is a transition from the primitive bacteria to the mushroom, where, apparently, that branch of the evolutionary tree ended. The blue-green algae seem to be at the end of another branch. Most marine bacteria are motile and possess flagella, though the flagella of bacteria are such fragile organs that their existence was at one time doubted. The tiniest flagellates are within the same size range as the bacteria, but their flagella are more developed and much stouter. There could thus easily be a connection between the bacteria and these small flagellates, and thence, when they began to manufacture chlorophyll, to the coloured flagellates discussed earlier; these too are similar in many ways to the spores of higher plants.

The most primitive organisms to exhibit photosynthesis are the purple and green sulphur bacteria. These are similar in form to other colourless bacteria of the marine world, but possess a primitive form of chlorophyll, resembling in its general structure the chlorophyll of higher plants. One can follow too, the progression through the flagellates to diatoms and other forms which have motile, flagellate spores at some stage in their development. More complex forms become colonial, and division of function between the individual cells leads to the development of multicellular organisms. Such forms as *Volvox,* formed of a number of independent, flagellate cells embedded in a spherical matrix, show the first stage, and the larger seaweeds with inner and outer sheaths show the second. Some of the colonial ciliates, which consist of a number of cells, each in a non-living sheath, are attached centrally to the substrate by a long stalk. The stalk slowly curls up, drawing the animals inward, and then each stalk suddenly and rhythmically extends to its full length, all at the same time, in what appears to be perfect coordination. And yet there is no physical attachment between the cells. We get the same type of movement in a diatom called *Bacillaria paxillifer,* in which all the cells behave in a like manner, moving parallel to one another, out and in, forming now a palisade, now an extended V or W (see page 47). This is one of the many rhythms of nature, developed even in primitive forms. The succession of species which we have studied, in marine environments, is at least partly due to such rhythms, as well as the consumption of certain nutrients and the production of antibiotic or growth-promoting substances. For there is evidence that certain microbes produce substances which inhibit or prevent the growth of competing species, but later produce self-inhibitory substances which cause the decline of the original species.

It is not always the weak species that disappears; sometimes the most dominant and aggressive species dies out, as man may do, because, having taken over the environment, it eats more than the environment can supply. The survival of a species is not always actually a case of the survival of the fittest, but of live and let live. Each organism has to maintain its own population, but must also relate to the community, as it ultimately depends on that community for survival. A thriving parasite is the one that has come to terms with its host. If it kills the host quickly, it also will die. The red tide organisms take over the environment completely for a short time, exhaust it and die out. However, their spores manage to survive in pockets or small areas in the sea, and can drift into areas where new outbreaks are possible. The late Laurence Baas Becking called this live and let live principle 'metabiosis' or 'living together'. It was his thesis that no organism could live alone, but always needed help in some form or another from even the most obscure member of the community. Thus, the continued existence of any species depends not on its degree of aggression, but on its ability to live with the other members of the community.

We have outlined the possible evolutionary processes from our knowledge of the form and function of microbes of the present day, and we know that present microbial behaviour was identical with or similar to that in earliest times. There is evidence that any changes that have taken place have been slow by our concept of time. As an example, bacterial activities of the sulphur cycle have been traced back for millions of years by modern isotope techniques. The sulphur bacteria sort out sulphur 32 from its isotope sulphur 34 in different ratios, and such ratios have been found in archaic sulphur deposits in which bacteria are believed to have taken part. There are also similarities in deposits derived from organic detritus, suggesting that the saprophytic bacteria of olden times did the same things that they do today. Dr. Wilhelm Schwartz has even found fossil bacteria and further evidence for their existence in old rocks, so we now have definite proof of the presence of bacteria at a very early stage of earth history.

It is clear also that the photosynthetic plants came into being relatively early in the history of life on this planet. The production of the organized part of the 'Weltschlamm' would diminish rapidly as oxygen was produced, probably by chemical means, and so the biological system would have run down rapidly if plants had not been around to supplement it by producing more oxygen from photosynthesis and more organic material from the sun's light energy. It is significant that the purple and green sulphur bacteria were present before free oxygen was available, and that these had the precursors of chlorophyll. It is but a short step from here to the blue-green algae which may also have been present at this time, or shortly after, and adapted to normal photosynthesis, producing oxygen from the reduction of carbon dioxide. Like the photosynthetic bacteria, the blue-green algae have no chloroplasts and contain chlorophyll *a* (the photoactive one) only, but no chlorophyll *b* or *c* as the higher plants do. They can also live in environments that are unfavourable to all but the bacteria, and seem to have a number of different means of assimilation: photosynthesis using carbon dioxide, photoassimilation using simple organic substances usually available in the water, and possibly photoreduction using hydrogen sulphide as a source of hydrogen as do the photosynthetic bacteria. The hydrogen is used to combine with the oxygen from carbon dioxide forming water and an organic molecule.

At this stage, respiration as we know it became possible and the aerobic life came into being. Plants would have to precede oxygen-using life forms, such as the aerobic bacteria. It is interesting that the primitive bacteria have been able to survive through the suceeding eras of earth history and that their chemical and biological processes are as necessary now as they were at the dawn of life. They were not a transient phase in the history of the world, but a permanent basis on which life was founded. Moreover, life is based on certain processes which have not changed over the millennia, but which have in higher organisms been partitioned out among cells which are united in space but differentiated in function. Thus, in plants we have the protective tissues of the leaf, the photosynthetic palisade tissue inside and the transport system and storage cells, all derived by changes in morphology of originally identical cells. In higher animals too we find the same differentiation, and even cells in the digestive tract and elsewhere which resemble the ciliates in form, and to a large extent in function. Each microscopic plant can perform all the same functions for life as a giant redwood or a field of grass.

**2. Microbes in Geology**

We have suggested that microbes, similar to or identical with many of those existing roday, were present in very early times, and that they played a role very similar to their present one. We have yet to discuss these roles in some detail. Firstly, let us consider the importance of the organisms themselves as contributing to the actual crust of the earth in the form of deposits. These deposits consist of organic materials such as detritus – the partly decomposed remains of plants and animals – silica and carbonaceous or limestone deposits.

Detrital deposits may be composed of slightly altered plant or animal material such as the fossil ferns associated with the coal measures, derived from freshwater swamps rather than marine environments; there are also plant remains associated with estuarine sediments. The laying down of slates and shales is influenced by organic detritus, the particles being frequently coated with organic materials and the rate of deposition being often controlled by these coatings, as in montmorillonites. Microbes produce abundant colloidal material, some up to 50 per cent of the total material produced by photosynthesis. Such substances are very important in determining the structure and precipitation of clays, and in the formation of sedimentary rocks, especially in estuaries.

The Texas Bays, particularly the Laguna Madre between Corpus Christi and the Mexican border (Fig.15), give some highly interesting and instructive examples of the importance of detritus in sediment formation. The sea-grasses *Diplanthera* and *Thalassia* are abundant in the area and produce large amounts of detritus annually. They die off at the end of the summer and drift into the shallows where they pile up. Bacteria and fungi degrade them and produce a sulphur cycle (see page 106). Usually, a layer of sand is laid down by wind or flood action and a series of layers of detrital material and sand will eventually compact into a sedimentary rock, probably slate or shale. An effect of the sulphur cycle and the organic matter is to produce sulphate in large quantities in the sediment, and this combines with the calcium in the water to form gypsum crystals which occur in large rosettes. This process is similar to that used for the commercial production of sulphur and sulphate by the action of sulphur bacteria on sewage. Joining in this process are the blue-green algae which usually form a carpet over the detrital material in the Texas Bays and help to separate the sandy layers, while also contributing extensively to the detritus. They also produce calcium carbonate which helps to bind the sediment by forming a cement.

It is generally believed that petroleum deposits are formed in surface or near-surface sediments in marine conditions, i.e. under estuarine or coastal influence. Dr. Brongersma-Sanders published a paper which has not received the attention it deserved.[1] She was studying what is known to marine scientists as the Walvis Bay phenomenon. This is essentially a red tide, which is endemic in that area of the west coast of South Africa, where there is also a very rich plankton and fish population produced by upwelling. The red tide organisms cause a mass kill of all living matter in the shallow waters of the bay, firstly by toxins and secondly by exhausting all the oxygen in the water and sediments, aided by sulphate-reducing bacteria and other organisms of the sulphur cycle. Initially, a single species causes the red tide, followed by many species and groups of organisms – dinoflagellates, bacteria, protozoa of many kinds, and possibly other organisms – which cause mass pollution. In the end, all animals and plants which require oxygen, including the abundant fish, are dead, and the bacterial flora takes over. When oxygen is absent, the only reactions possible are those which end in the addition of hydrogen, such as the formation of ammonia from nitrate, methane from carbon dioxide, or sulphides from all sulphur compounds. It is quite conceivable that under these conditions, the complex carbon compounds from the animals and plants will be converted into compounds consisting only of carbon and hydrogen, the nitrogen escaping as ammonia and the sulphur as hydrogen sulphide. If conditions were suitable, and Dr. Sanders thought they were, these carbon-hydrogen compounds would be those which form the basis of petroleum. The somewhat limited pressure of deposits above them would probably give the necessary factors to change some of these hydrocarbons into the ones that are normally associated with petroleum as we know it, and with the natural gases. It is significant that sulphur in one or more of its compounds, or as natural sulphur, is usually associated with petroleum. That is what we would expect in a formation of this type. Many people have tried to make petroleum using bacteria, particularly the sulphur bacteria, but none have succeeded. Bacteria produce the sort of reaction that could lead to petroleum formation, but not the critical reactions themselves. I have always wondered why the oil companies have not followed up Dr. Sanders' suggestion and tried to create an artificial Walvis Bay phenomenon in a laboratory or small field plant. I believe that oil formation is a complex phenomenon requiring the cooperation of a number of micro-organisms, probably coupled with the presence in the substrate of many larger organisms, to provide the necessary organic complex from which the various hydrocarbon entities may be derived. The relative simplicity of the composition of petroleum, as against the suggested mixture from which it may come, is due to the simple products of bacterial transformations under strictly anaerobic conditions. It is known that oil deposits are not found in the areas in which they are formed, but flow or are squeezed into the tops of anticlines or into porous strata. This movement would itself serve as a separatory funnel to remove the petroleum from the other less moveable organic constituents. Nitrogenous compounds would be lost during these processes because all inorganic compounds of nitrogen are highly soluble in water, and would be easily washed out at any stage in the process.

An important role of microbes in geology is in the formation of sediments. Radiolarians form tiny shells or skeletons of silica, which may settle on the sea floor. For example, in the vicinity of the Jenolan caves of New South Wales, Australia, a rock stratum over 1,000 feet thick consists primarily of shells of radiolaria; it is known as the radiolarian jaspers of the Jenolan Series. Either the organisms must have been growing at an incredible rate compared with their present growth rates, or they must have been living in a very stable environment for an exceedingly long time. Radiolaria are not very common in plankton, which means that there has been a great change in the abundance of different types of microbes during geological ages. The same can be said about another group of microbes, the silicoflagellates, which rarely occur nowadays in any numbers except in certain arctic areas such as Peril Strait and Sitka Bay in Alaska. The diatoms also provide siliceous skeletons for diatomaceous oozes, but some of these such as the well-known ones at Oamaru in New Zealand were formed in freshwater lakes. There are, however, diatomaceous deposits in places such as the Challenger Deep, in this case composed largely of *Coscinodiscus rex,* a species which is very large (about 1 millimetre in diameter and the same depth), and which is very rare nowadays. It has been recorded in some numbers off the coast of California, but I have not found a living specimen in any of my collections. There is a band of diatomaceous ooze forming in the Antarctic, and this dates from the pleistocene to the present day. Calcareous deposits which form geological strata may be due to the coccolithophores and the foraminifera, most of which have calcareous skeletons, though one species uses strontium instead of calcium.

Foraminiferous deposits in the oceans or oozes (see page 33) are usually ascribed to the sinking from the surface or near-surface of thousands of these tiny organisms. This may happen in shallow waters and enclosed bays. It does not happen in deeper waters such as the Antarctic deposits, where it can be shown that living diatoms are carried downward with the currents, and that they can live without light for a considerable time.[2] Sediment with this diatomaceous ooze is composed in part of partially dissolved fossil species, which are no longer found alive, and recent species from the waters above. The skeletons of these organisms are always fresh-looking and intact. This would not be the case if they had drifted slowly down through the various water layers, and anyway they would have been carried far to the north by one of the lower water currents; the species found in the deeps or in diatomaceous earths are mainly those which are to be found in shallow sediments or at the surface today. Because of the slowness of the sinking of small particles through the water, the possibility of the solution of their hard parts and the vagaries of water currents in the ocean, it would be impossible to produce a concentration of diatom or other skeletons in any one area on the scale of our fossil deposits. They would be well scattered, with perhaps some local concentrations but not definite bands. Fossil beds such as the miocene Californian deposits contain mostly bottom-living diatoms.

I have mentioned that certain species of diatoms which are to be found in fossil beds do not occur today. This applies also to other microbes, and allows them to be used to determine with some degree of accuracy, the age of the stratum. Some species were decidedly ephemeral, while others persisted through many geological ages. The reason for this is not clear, but the phenomenon is very useful to the geologist who can often use it to determine not only the age, but also under what conditions the materials were laid down.

It is interesting that Dr. Schwartz and others have found what appear to be fossil bacteria, since these microbes have no hard parts. Sometimes, of course, fragile organisms may have parts replaced by silica or lime as in the case of the petrified forests. The ciliates do not seem to have left any mark of their occurrence in fossil beds, but besides the calcareous foraminifera and coccolithophores, the dinoflagellates do occur as fossils due to the refractory nature of their skeletons. Palaeontologists who study them however do not try to homologize them with present-day forms, but use a different classification. I am not sure how much justification there is for this, and it might be interesting to try to make detailed comparisons assuming close relationships. We might be able to determine the lines of descent of some present-day forms, and thus make a guess as to the relationships between genera and species.

In tropical areas, we find huge deposits of coralline limestone, together with rocks known as 'oolites', also composed of calcium carbonate. The South Florida-Bahamas region is typical of this sort of formation. Corals build up slowly as the shoreline is sinking, and die when it rises above the water level. We have seen that the coral mass is controlled by the activity of the primary producers, the coralline algae and the zooxanthellae, together with some phytoplankton. So coral limestones could not be formed except for the microbes of the reef. The activity of microbes in coral reefs is not confined to these primary producers.

On a coral reef there are large tunnels, deep and shallow channels and a lot of smaller crevices. In a coralline limestone, however, the whole is cemented together with the coral heads intact, and the interspaces filled with broken coral and a more or less amorphous matrix. Some of it is made of powdered coral broken by the waves pounding on the reef, but some is produced by microbial action. In a marine system, the animals and plants change the acidity and alkalinity according to a diurnal pattern by respiration and photosynthesis. In limestone country, this is important, as calcium carbonate is very easily dissolved and reprecipitated. When plants are using carbon dioxide or animals and plants are producing it, there will be a solution or precipitation of calcium carbonate, i.e. the dissolving or formation of limestone or chalk. Thus, solution and precipitation are produced in the same reaction, according to rather slight changes in the conditions, and these conditions are easily altered by living organisms, especially the microbes. We have learned that the plants change the pH easily, but the bacteria will do this also by producing such substances as tannins and other organic acids, such as acetic or lactic. They can also produce alkaline conditions, as by denitrification, i.e. the reduction of nitrate to ammonia; in fact, it has been suggested that denitrifying bacteria are important in the solidification of coral reefs. The production of hydrogen sulphide or sulphur from sulphates will also raise the pH, and detritus-degrading bacteria may be of great importance as they exist between the coral heads where the matrix is formed. This phenomenon of solution and reprecipitation occurs in limestone country as one may see in limestone caves, and is especially active in flat areas such as the Bahamas and South Florida. The reprecipitated chalk, which may be transported some distance from where it was dissolved, often forms the kind of material known as oolite, where the precipitation occurs around particles of plant or other material, especially the blue-green algae. These dissolve their way through limestone and at the same time form a network of tiny strands of chalk around their own threads, and these compact in the formation of the oolites and calcareous sands. Thus the powdery formations in the interior of coral atolls and lagoons are formed.

Microscopic algae, especially the blue-greens, cause not only the precipitation of chalk, but also may act in conjunction with the bacteria in the formation and alteration of calcareous rocks such as gypsum and dolomite, which contains magnesium as well as calcium. In the production of carbonate rocks, as well as in their solution, microbial activity is important; though there is a great deal more to be discovered about it.

You may wonder about the presence and activity of microbes through the geological ages. Bacterial and algal activity has been demonstrated from the Proterozoic – that is the pre-Palaeozoic – age, the dawn of life. Thus, our theory of the origin of life is supported by geological observation. The foraminifera were present in the oldest Palaeozoic rocks and the radiolaria appeared around the time of the coal measures of Britain and the United States, while coccolithophores and dinoflagellates did not appear until later, followed by the silicoflagellates. The diatoms appeared much later, and certain species characterize specific geological strata (see page 117). Diatoms are also being used to indicate the origin of sediments in the Black Sea and the Atlantic Ocean. Different species are characteristic of fresh, brackish and salt water, and can be used with some accuracy in estuaries to show salinity gradients. The freshwater species can be carried into the ocean by streams and currents, for example by the waters of the Congo or the Niger, and then moved in one direction or another by ocean currents. There are certain species which are characteristic of the Amazon delta and these can be followed with the movement of the Amazon water as it is pushed to the westward by the South Equatorial Current, and show mixing of Amazon and Equatorial waters as they persist for a time in the mixed water. Diatoms that were growing in areas such as the Sahara desert when it was lush countryside may be and are transported by wind across the Atlantic, and have been caught by wind nets on the eastern shore of Barbados. Even living microbes are caught up in winds from the surface of the sea, and transported for many miles. Living microbes can even be transported across the Isthmus of Panama, and this probably accounts for the similarity of the phytoplankton on both sides of this apparent barrier.

The aerial transport of diatoms is believed, on present knowledge, to have been important in forming the sediments of the mid-Atlantic ridge.

I earlier discussed the various chemical processes which are or may be catalysed by marine bacteria. It seems certain that bacteria are responsible for the production of mineral sulphur such as occurs in the sulphur domes of Louisiana. Doctors Butlin and Postgate of Britain made a study of some lakes in Cyrenaica where sulphur is thus deposited and is harvested locally.[3] Gypsum (calcium sulphate) supplies the sulphate for the bacteria, which change it to sulphide, and the purple sulphur bacteria, which we have previously discussed change the sulphides back to yellow sulphur on the surface of these lakes. In parts of Great South Bay, Long Island, large quantities of a seaweed, *Cladophora,* grow abundantly, die at the end of the summer and are washed into the boat channels. There they rot and bacteria produce hydrogen sulphide in such quantities that it carries large blobs of the weed back to the surface of the water. Here the sulphide is oxidized by bacterial action or chemically in contact with the air, and the water turns milky with tiny particles of sulphur.

The bacteria of the sulphur cycle are also responsible for restoring phosphate from the sediment to the water, and this prevents the build-up of phosphate in wet sediments.

Microbes can concentrate a number of elements up to 10,000-fold, and bromine, iodine, strontium, vanadium, potassium and boron are among the substances that can be concentrated in this way. The formation of bacterial sulphide is primarily responsible for the deposition of iron minerals such as pyrite, but chemical changes can alter pyrite to copper pyrites and other sulphide minerals such as galena (lead sulphide) may be formed in this way. Even uranium ores have been considered to be of bacterial origin, but evidence for this is difficult to obtain. In general, microbes play a very important part in a number of geological processes, but they are not magical in their properties or activities and cannot perform reactions which are not possible otherwise. They can accelerate a reaction, or maybe change the place where it occurs by carrying the components or concentrating enzymes in certain areas.

## NOTES


1. Brongersma-Sanders, K. 'The importance of upwelling water to vertebrate palaeontology and oil geology.' Verh. Koninkl. Ned. Akad. Weterschaps. Afd Natuurk. I *45* (4): 1-112.
2. Hart, T. J. 1934. Phytoplankton. 'Discovery' Rept, *8*: 188-9.
3. Butlin, K. R., and Postgate, J. R. 1954. 'The Microbiology of sulphur of Cyrenaican lakes.' In *Biology of Deserts.* Inst. Biol., London: 112.

# PART V

# MARINE LIFE AND MAN

# 14. MEN AS DESTROYERS OF MARINE LIFE

Man is horrified by the thought of a nuclear war, but careless of the fact that he is at war with nature. The Australian aborigines, men of the stone age until very recently, were never a threat to the environment, because their population, prior to the coming of the white man, was controlled by nature at a level of possibly 500,000 in a continent of about 3 million square miles. Other stone age populations in other parts of the world were similarly controlled, and this, coupled with environment control of other predators, kept a balanced ecology. The change came when man started to husband his resources, to cultivate the land, fell the forests and dam the rivers and streams. Not only did he create an artificial environment, but he also found that he had the power to divert nature and interfere with natural phenomena. From then on he became increasingly destructive, and can now easily turn the whole earth into a desert. He already has done so in the Middle East, in the Sahara, and in parts of Australia.

Man's attack on the ocean is still recent. But even the Japanese, the world's most intensive fishermen, now have to range the world to keep up their supply of fish. They tend to capture and eat everything that is present in their nets, and so upset the ecological balance less than the fishermen of the western countries, who are highly selective in the fish they will catch.

It was at one time believed that heavy fishing gears, such as trawls dragged along the sea-bottom, ploughed over and damaged the soil and the creatures living and breeding in and on it. This is not nowadays generally regarded as a potent source of damage to the fisheries, though there is one case where there seems to have been real damage. Along the coast of New South Wales, Australia, a fishing ground known as the Botany at one time supported twelve trawlers; but now it cannot support any. There is a strong 2 to 3-knot current flowing south along the coast, and the continental shelf is narrow, only 14 to 20 miles wide. The muds stirred up by the trawls and their warps were carried by the current over the edge of the shelf into deep water, and thus lost to the inshore environment. As the fish were largely dependant on the flora and fauna of this shelf, they disappeared too. The sea-bottom was originally a silty mud with a rich bacterial flora and many diatoms, flagellates and ciliates; it is now a pure sand and the silt has disappeared – since silica sand is heavier than silt, it was re-deposited when the silt was drifted away. But silt has a much richer microbial count than sand, where the content of organic matter is almost nil. With the microbes went the detrital material, and thus the basic components of the food-web were removed, and this ended the fishery.

Research in fresh waters, and in paddy-fields under irrigation, has shown that the sediments clearly act as a storehouse and regulatory mechanism for nutrients such as nitrogen and phosphate; this is largely true for estuaries also. I have found that there is a large population of microscopic algae, bacteria, and protozoa in the surface sediments of both estuaries and coastal shelves, and it can also be shown that there is a large and steady population of crustaceans feeding on these microbes. The short life-cycle of the microbes means that there is a rapid turnover of nutrients. Dr. Johannes of Georgia[1] has found that the rate of turnover of phosphate, that is, the time between intake and excretion, is inversely proportional to the size of the organism; so it is an advantage to have a large number of microbes in the ecosystem rather than a smaller number of large animals or plants. In shallow and relatively shallow waters, the small plants, which are attached to other plants or objects or are living attached to the bottom, are the important primary producers; they are thus important to us in our studies of the productivity of such areas. I have never found, except in the pilchards and sardines and some of their allies such as the Peruvian anchoveta, any great number of planktonic plant cells in the stomachs of fish, though many shellfish such as mussels and oysters feed on phytoplankton, including many very minute forms. The Toheroa of New Zealand lives on a peculiar and very abundant plankton diatom called *Chaetoceros armata,* which has a very strange distribution, namely the east coast of Britain and the west coast of the north island of New Zealand.

Pollution in estuaries is exceedingly important to the microbial population, and a number of changes can be attributed to this. The red tides of the eastern Gulf of Mexico have been ascribed to the eutrophication of the Gulf waters by fertilisers, and such crop by-products as vitamin $B_{12}$ and especially phosphate from the extensively-mined phosphate deposits of western Florida. This eutrophication is attributable to the farming practices east of the Mississippi, and subsequent run-off from the land. Under some conditions sewage pollution can produce or intensify local anaerobiosis and consequent increase in the production of hydrogen sulphide by bacteria; this may produce devastation close to the source of the discharge, but beyond may be a strong fertiliser to nourish an intense growth of microbial epiphytes and plants of the sediments such as diatoms and flagellates. These may enhance the local fish and shrimp populations, or, alternatively, cause over-fertility, resulting in the growth of weeds, or algae supporting an unwanted population. It seems that careful planning for efficient use of the effluent, as indeed is done in many sewage works, could ensure benefits in terms of increased fish and shrimp production. The production of sulphide wastes from paper mills causes the death of plants and animals near the outfalls; but at greater dilution, as in Sitka Sound, may nourish a greater abundance of microscopic planktonic algae than in the environment generally.

Recently it has been suggested that the world's oxygen balance might some day be threatened if the production of oxygen by the phytoplankton of the oceans were curtailed by pollution. This is false. We have enough real pollution problems facing us in the sea without tilting at windmills.

First, even if we set out systematically to destroy the phytoplankton of the oceans it is quite unlikely that we could do so. It is true that pollution in some coastal areas has resulted in the destruction of phytoplankton production locally. But pollution great enough to do this on a global scale is very hard to imagine.

Secondly, even if we did destroy the world's marine plant life, it would have virtually no effect at all on the world's oxygen balance. The oceans' phytoplankton produce in a year only about 1/10,000th of the amount of oxygen already present in the atmosphere. Furthermore virtually all of the oxygen that is produced in the sea is consumed there by marine microbes and animals. Thus marine productivity makes no net contribution to the world's supply of atmospheric oxygen. There are innumerable ways in which we can make our world less habitable for our descendants before we make a small dent in our oxygen supply.

Marine microbes are the most potent detoxifying agents in the sea. Left to themselves, oil spillages for example are to a large degree 'cleaned up' by marine microbes. After the lighter fractions of the oil evaporate, the heavier fractions are attacked by bacteria and other microbes. Finally the solid residues, when cast ashore, are devoured by many browsing molluscs. But this purification takes time; and in the case of the wreck of the supertanker 'Torrey Canyon', for example, massive oil discharges were drifting about at the beginning of the holiday season off the coast of Cornwall, where tourism is the main industry. Hence an immediate campaign of detergent-spraying, well played-up by the media, was begun. These detergents, it was soon discovered, did more damage to the environment than the oil itself.[2]

The natural bacterial control of putrescible and other discharges can contribute powerfully to the fertility of sea and fresh waters. It is said that the sewage discharged into Manila Bay in the Philippines enriches the water to the extent that the fish crops from the marine fishponds around the bay are improved. On land, sewage is exposed to natural bacterial purification in sewage farms, where the sludge is spread out so that the maximum surface is exposed to bacterial action. In Munich and many other cities, anaerobic decomposition of the sludge produces methane, which is sold to enrich town gas;[3] the residual solids are sold as fertilizer, or as compost-makers under proprietary names. Much of it is pumped to silos from which it is used for the reclamation of moorland. The liquid is pumped into large fishponds, where it slowly oxidizes, releasing nutrients which result in very large fish crops. All these processes earn money, so the treatment of sewage in these cases, thanks to anaerobic and aerobic microbes, almost pays for itself and is a relief to the ratepayers. The brackish-water fishponds in the delta of the Ganges benefit greatly from the discharge of the Calcutta sewage works.[4]

In Guyanilla Bay in Puerto Rico there were discharges of volatile and other oil fractions from a petrochemical plant, and also from a sewer at Ponce. Purification of the oil residues began in the outfall channel, and microbiological activity decomposed the hydrocarbons and fostered the growth of algae which oxygenated the water. The sewage was purified by a different group of micro-organisms. Both processes contributed to the enrichment of the water of the bay.

Certain microbes can assimilate and concentrate inorganic ions; and it may be possible to use such organisms to reduce the toxicity of wastes containing metals such as copper, iron, and mercury, and even strontium and vanadium.

A study of the heated outflow of cooling water from a power plant at Turkey Point on Biscayne Bay, Florida showed that in winter, the growth of the attached algae was increased, but in the summer, it was reduced; the number of species occurring was diminished. Unfortunately, the epiphytic algae had no natural materials to attach to – only the artificial materials supplied, so we can hardly say that the warm waters in winter made the environment more suitable for the fisheries of the area. In the sediments near Turkey Point, the diatom-flagellate flora was replaced by a blue-green algal mat, and this, as we have seen, is a sign of stress, and of deterioration of the environment. The fact that improvement in the conditions for growth of microbes occurs in some cases and under certain conditions in areas regarded as polluted surely is an indication that we should make further and more detailed studies of the conditions for optimal growth of the micro-flora to see whether we can take advantage of estuarine pollution.

## NOTES


1. Johannes, R.E. 'Uptake and Release of dissolved organic Phosphorus by representatives of a coastal marine ecosystem. *Limn. Oceanogr. 9*:235-242.
2. Smith, J.E. (Ed.). 1968. *'Torrey Canyon' Pollution and Marine Life.* Cambridge, at the University Press.
3. Scheuring, D. L. 1936. 'Die Reinigung and Verwertung der Abwasser von München.' *Natur und Volk 69*: 390.
4. Nair, K. K. 1944. *Calcutta Sewage Irrigation Fisheries.* Proc. Nat. Inst. Sci. India. *10*: 459-462.

# 15. DESTRUCTION OF MAN-MADE MATERIALS BY MARINE MICROBES

An important aspect of microbial activity in the sea, at least from the economic point of view, has to do with the corrosion and destruction of man-made structures, including ships, steel and concrete pilings, rope, wharf pilings and similar things.[1] Bacteria are believed to form a primary film on steel ships; this coating prevents or delays the copper or mercury of the antifouling paint from keeping off the larger fouling organisms such as barnacles or tube-worms. Dr. Hendey, in Britain, has found that some diatoms have a higher resistance than others to both copper and mercury, the poisons actually used in antifouling paints for the control of growths or ships bottoms.[2] Microbes of several kinds can, therefore, reduce the effectiveness of antifouling paints and shorten their active life. They also play a part in the destruction of antifouling and anti-corrosive paints used on the hulls of ships, mainly by attacking the vehicle – the part of the paint that causes it to flow and to adhere to the surface. Very important in the corrosion of hulls and steel structures are the bacteria which reduce sulphates to sulphides. Sometimes a blister in the metal of a hull, or a small 'holiday' or gap in the paint film, will allow bacteria to collect against the steel, while the decomposition of organic detritus sticking to the hull removes most of the oxygen. As a result, a small area has a lower electrode potential than the rest of the hull, that is a lower redox potential, and a small electric current passes from one area to another; part of the metal then goes into solution by electrolysis. The reaction spreads, the paint sloughs off, and cracks appear, allowing further attack by the bacteria. When the mothball fleet of the United States Navy in San Diego harbour was activated for the Korean war, it was found that this kind of corrosion had occurred on a number of ships, just below the waterline and all around the hulls, so there was a band of partly removed and altered metal right round the ship. Considerable repairs were necessary before the fleet could put to sea.

Corrosion of steel pilings at the waterline in salt water is another nightmare of marine departments, and this is also due to the bacteria of the sulphur cycle. Corrosion of concrete is another serious problem, and a bacterium has actually been named *Thiobacillus concretivorus* since it 'eats' concrete. These microbes not only attack the materials used in man-made structures in the sea, but also paints and other means used to protect them.

Wooden structures are attacked by fungi, many of which are stated to be indigenous in the sea. Many studies, mainly taxonomic, have been made of these microbes. A number of marine bacteria and fungi are capable of attacking and destroying lignin and cellulose and thus breaking down the structure of wood and cordage. It is believed that bacteria and fungi also assist marine borers in destroying wharf and ship's timbers. Fungal destruction of wharf timbers, especially between wind and water, can be quite serious and is often insidious. All will appear structurally sound until a ship bumps the wharf and the affected timbers give way. Usually exposed timbers are impregnated with creosote or a copper salt so as to delay or prevent fungal attack. Cordage too is attacked by fungi, many of them of marine origin, though some are present in the rope as it is woven. I have seen a large (8-inch) mooring-rope break during the mooring of a ship, due to the destruction of the centre of the rope by marine fungi. This rope had been tested before the departure of the ship from home port, but had been almost continuously wet between then and our arrival at the next port, so the rotting had been able to go on unnoticed and unchecked. Fishing nets are attacked by microbes, and serious losses can be sustained by fishermen if their nets are not treated and are left stacked while wet; though nowadays most nets are made of nylon or other plastics.

Marine microbes are also sometimes parasites of fish and other marine animals of commerce, and cause what is known as 'fish spoilage', the deterioration of fish between the time it is caught and when it reaches the consumer. The barracouta (or 'snoek') is sold in the markets of southern Australia, but the fish is often badly affected by a protozoan parasite which turns the flesh milky. The whole flesh breaks down soon after the fish dies, until it oozes out like tooth paste when the body is squeezed. As the process goes on even at low temperatures, the disease has made the fish unpopular with both the retailer and the consumer. Other microbial diseases that affect the marketing of fish are due to a bacterium that attacks the shell of lobsters and crayfish, and another that lives in the gills of mullet in tidal areas and gives an earthy taint to the flesh. Microbial parasites in marine animals are unexpectedly rare, however, because fish range over wide areas, and diseased fish will have difficulty in keeping up with the school.

The production of poisons by microbes, particularly certain dinoflagellates, may either kill the marine animals as in Walvis Bay and the eastern Gulf of Mexico, or may produce poisons which are ingested but do not harm the animal. This happens in the shellfish poisoning of clams and oysters in the bays of Washington and Oregon states, on the west coast of the United States. In this case, a nerve poison is produced by the microbes, stored by the shellfish, and is quite capable of killing humans who eat the clams. Similar shellfish poisoning has been reported from Canada and Europe, but only on rare occasions. In certain areas such as the Myall Lakes in eastern Australia, one may find a bacterium (*Staphylococcus*) related to the organisms which cause blood-poisoning and hospital infections. This particular strain is able to live in marine environments, and may grow and multiply in badly stored shrimp and lobsters, causing severe poisoning to people who eat the meat even after it is cooked. Both the bacteria and the poison are tolerant of heat, and can survive the average temperatures involved in cooking fish. Every gourmet knows that fish should be lightly cooked so as to conserve the flavour and texture.

The most deadly poison known to man is that of a bacterium found in the sea – botulinus toxin – and produced by an anaerobe, *Clostridium botulinum.* This is the toxin which has been prepared for use in bacteriological warfare, and which, if dropped in reservoirs or even over cities, could destroy whole populations, and is probably more devastating in terms of human life than the hydrogen bomb. This toxin is even more resistant to heat than the *Staphylococcus* toxin, and can still be potent in canned foods if they are not correctly processed. Fish canners try to use as little heat as possible, because the less a fish product is heated the better the flavour. Japanese canned crab is whiter and has more flavour than Alaskan canned crab because there is no botulism in Japan so lower processing temperatures can be used.

It is well known that fish begins to deteriorate quite rapidly after it is caught, and that it should be iced or frozen as soon as possible. Part of the deterioration is due to the release of enzymes from the tissues of the fish, and these attack the parts they previously protected. The major part of the deterioration is due to various bacteria, most of them originating in the water or the sediment, many being found in the slime which occurs on the surface of the bony fishes, but not on the sharks or rays. Other bacteria occur in the gills and, when the fish have been feeding, in the gut. In some parts of the world, fishermen starve captured fish for some 48 hours before they kill them for market. Fish treated in this way will last longer than fish packed immediately after they are caught. In parts of South Australia where fishermen starve their fish before packing, they last without noticeable deterioration for over a week when packed in ice. Fish not so treated, and sent from ports closer to the market, show deterioration within 36 hours. It can be seen that the bacteria from the gut of fish are an important source of spoilage and deterioration in market fish.

In fish spoilage, the bacteria seem to follow a certain sequence; first yellow and red forms, then white ones, and after that, bacteria which produce a green pigment which soaks through the flesh. Actually, none of these bacteria are poisonous except an occasional strain with the green colour, so really badly decomposed fish can ordinarily be eaten. Some of the eastern peoples actually prefer their fish in this state. Often, in fish canneries, the flesh is allowed to spoil slightly as the enzyme action on the flesh tenderizes it and makes the pack more palatable to the consumer – provided that he doesn't know it. It is very difficult for an inspector to say that fish is unfit for human consumption, as people can and do eat rotten fish without bad effects.

## NOTES

1. Firth, Frank E. (Ed.). 1969. Encyclopedia of Marine Resources. Van Nostrand Reinhold Co., New York: 417-18.
2. Hendey, N. I. 1947. Copper in diatoms. *Nature 159*: 646.
   Hendey, N. I. 1951. 'Littered diatoms of Chichester Harbour with special reference to fouling.' *J. R. Microspecial Soc.*: 1-86.

# 16. MEASURING, STIMULATING AND UTILIZING MARINE PRODUCTIVITY

A variety of methods have been tried for estimating primary production, or annual production of plant material in the sea. Dr. Harvey of Plymouth suggested that as all plants contain chlorophyll and animals do not, measuring chlorophyll would be a good way of estimating phytoplankton abundance.[1] (Phytoplankton cannot be physically separated from zooplankton and weighed.) He extracted the chlorophyll from water samples and measured it in an ingenious instrument he had designed himself. However, the assumption that all the marine plants contain the same amount of chlorophyll relative to the weight of living material in them cannot really be accepted, so the method gives only approximate results.

To measure the growth rate of phytoplankton, water samples are sometimes put into glass bottles, one in the light, one in the dark, and changes in oxygen levels in the bottles measured over a period of time. Since oxygen is consumed in respiration and produced in photosynthesis, the difference between oxygen levels in dark and light bottles indicates the net amount of photosynthesis that has occurred and is thus a measure of plant production.[2] This method is rather insensitive however and works well only in productive water.

Dr. Steeman-Nielsen of Copenhagen devised a much more sensitive method of estimating growth rates by measuring the amount of carbon dioxide taken in by phytoplankton.[3] He collected his seawater samples, injected radioactive carbonate, and measured the amount removed from solution by means of a Geiger counter. This seemed for a time to give the required answers, but a number of sources of error have been found which are difficult or impossible to avoid. The results are again, therefore, only approximate.

TABLE I

ANNUAL PRODUCTION OF LIVING PLANTS (TONS PER ACRE)

| | |
|---|---|
| Open ocean | less than 3 |
| Continental shelf waters | 1 – 10 |
| Salt marshes, some estuaries, alluvial plains, modern agriculture | 30 – 80 |
| Deserts | less than 2 |

Counting the microbes has also been used, but once more it gives only approximate results. It does tell us to some extent the proportions of the principal kinds of plants, and as there is increasing evidence that many animals are very selective in feeding, this may be more important than the total amount of plant material. I found, for example, that in the Gulf of Guinea the diatom *Coscinodiscus* was apparently consumed, whereas *Rhizosolenia* was not. The tiny flagellate Coccolithophores were almost completely consumed, but as they reproduce very rapidly, they were replaced by the next afternoon.

The sea has often been portrayed in the past as holding vast food reserves for man. We know, in part because of our measurements of marine plant growth, that the truth is considerably less impressive. The sea is not very biologically productive relative to the land. Certain coastal marine areas such as salt marshes, mangrove swamps, seagrass communities, and coral reefs are among the most biologically productive natural communities anywhere on earth. But the *average* productivity of the oceans is low. In fact the deserts of the world are only slightly less productive than the open ocean (Table I).

Plant growth in the desert is limited largely by lack of water, whereas plant growth in the oceans is limited by the low levels of dissolved nutrient elements like phosphorus and nitrogen. Iron or silicon may also sometimes limit marine plant growth. Evidence for this has been obtained in the following way.

A seawater sample is divided into a number of equal portions. Different portions are 'fertilized' with different chemical elements suspected of limiting plant growth. All portions are then injected with radioactive carbonate as described above. Phytoplankton in the water sample which has been enriched with the element that was limiting plant growth will be stimulated by this addition. It will therefore grow faster and take up radioactive carbonate faster than phytoplankton in the other samples. It is in this manner that marine biologists identify which element is present in the lowest concentration relative to the plants' needs. This element is often the weakest link in the chain of environmental requirements for phytoplankton growth.

Nutrient-rich soil may contain thousands of times the levels of plant nutrients found in the equivalent volume of seawater. Plant growth can take place through a vertical range in the sea, because of its transparency, but this does not counterbalance the limiting effect of low nutrient levels on marine production. It has been calculated that while the oceans occupy seven-tenths of the earth's surface they account for only about half the earth's plant production.

This being the case it might seem reasonable for the oceans to supply about half man's food. We obtain only about 5% of our food from the oceans because we harvest our food further up the food-chain in the sea than we do on land. The higher up the food-chain we harvest, the smaller the fraction of total biological production we get.

Much of the food energy ingested by an animal does not go into its growth. Some is respired or burned up to provide the energy necessary to run life's vital processes. A proportion is unusable and goes right through the animal without being digested. Most of the energy of plant production is consequently not transformed into animal tissue. Only about 10% of the energy in any level in the food-chain ultimately finds its way into the living tissue of the level above it. Therefore, it we harvest plants we obtain about ten times as much food per unit area as if we harvested animals that fed on these plants (herbivores), about 100 times as much as if we harvested animals that feed on the herbivores (primary carnivores), about 1,000 times as much as if we harvested animals that fed on the primary carnivores (secondary carnivores), and so on up the food-chain.

On land, where our production of food is carefully controlled by the farmer, we take this diminution in energy along the food-chain into strict account. We eat either plants or herbivores. Furthermore, the poorer the country, the greater proportion of plant material in people's diets. Farmers can produce more food in the form of plants than of animals, so plant foods are cheaper. And because of the progressive loss of energy as it travels up the food-chain, herbivores in turn are much cheaper than carnivores. Carnivores are so expensive to produce that even the people of the richest countries do not eat carnivorous land animals to any significant extent.

Most of the oceans' crop of plants is in the form of the microscopic phytoplankton spread very thinly through the upper water column. It would be necessary to filter more than a million gallons of water on the average to harvest a pound of phytoplankton. There is accordingly no known way of harvesting these plants economically.

In addition, the herbivores in the ocean consist overwhelmingly of zooplankton which are likewise so small and spread so thinly, that with rare exceptions they too cannot be harvested economically.[4] It is doubtful if man would be willing to eat most types of zooplankton anyway. Although many zooplanktas are relatives of such esteemed seafoods as shrimp, lobster and crab, their chitinous external skeleton cannot be peeled off easily. The zooplankton mixtures I have tasted were evocative of salted plastic scouring pads. However, research continues.

The bulk of man's catch therefore consists of carnivorous fish such as salmon, herring, tuna and cod. As a consequence we are utilizing only a small fraction of the oceans' productivity. Many of the world's major fisheries are already being fully exploited (or overexploited). There is little hope that our traditional fisheries can expand more than two- or threefold – and certainly not fast enough to fill the ever-growing army of new mouths in the world.

Many people have been looking optimistically to mariculture, or farming of marine fish and shellfish, as a possible means of rapidly explanding the sea's yield of food to man. Oysters have been cultivated by man since Roman times. The Japanese farm oysters, shrimp, edible seaweed and a variety of other seafoods. Milkfish and mullet have been raised in brackish and marine ponds in south-east Asia and Polynesia for centuries. The French and Italians farm mussels and oysters. With the growing expectation that expansion of traditional fisheries cannot continue much longer, many additional countries have begun to experiment with new species and new techniques for marine farming.

Farming of the open ocean is not practical and mariculture must therefore be restricted largely to shallow coastal waters for the forseeable future. There are two reasons for this. First, most biologically productive marine waters are generally found along coastlines because plant nutrient levels in coastal water are usually much higher than those in oceanic water due to run-off of nutrients from the adjacent land. The marine farmer, like his terrestrial counterpart, naturally wants to focus his efforts on nutrient-rich systems.

Secondly, shallow coastal areas are the only regions of the ocean that are amenable to much control by man. The oceans consist of a multiple series of streams moving vertically and horizontally in certain general directions, which are beyond man's control.

The Suez Canal and the Aswan Dam's effect on the fisheries of the eastern Mediterranean are small examples of how wide are the implications of man's bright ambitions on the environment.

It is also assumed by the proponents of oceanic fish production that upwelling and mixing are the only factors necessary for a rich fishery. That most large fisheries of the world occur on shallow banks or close to a continental shelf seems to indicate that more than a mere upwelling is necessary. We know that deep water frequently has more inorganic nutrient salts such as nitrate and phosphate than surface water, because the animals have been using the organic material more quickly than the plants can replace it; in fact, below the zone of light penetration (the 'photic zone') there will be no replacement at all as far as we know. However, it is not certain by any means that inorganic salts alone provide fertility. A factor which is not usually considered in studies of fishery distributions is the relation between the shallow bank itself and the water above it. Many microbiological processes, mainly bacterial in origin, occur in the sediments, especially at the interface where the sediment meets the water. It is probable that a number of vital materials which are not available in the open sea or in deep waters are produced in this biologically active area. Its importance is hinted at by the difficulty, and often impossibility, of growing many of the microscopic plants of the ocean in any form of culture. The most active nutrient mixture seems to be what is known as 'earth extract', which is prepared by boiling garden soil. Its composition and the reason for its activity are still unknown, but despite these drawbacks it has helped us to culture a lot of species, though by no means all, and it is hard to replace it by known chemicals in what is called a 'defined medium'. We must conclude that our knowledge of the nutrition and food requirements of the microbes that we want to cultivate as primary food is quite inadequate, and that what success we may have will be based upon chance, with grave danger of not being able to reproduce our results.

Furthermore, although there are large tuna fisheries over deep water, most big fisheries of the world are associated with banks or shelves, and with the upward trend of waters from the deep to mix with the surface layers. Some plans for mariculture are concerned with mixing bottom water high in nutrients with surface water, in which most of the nutrients have been exhausted. The mixing is to be done by pumping or by the use of atomic power, putting a reactor on the sea-bottom and driving the bottom water upward by convection heating, or by other artificial means.

We may be able to produce artificial changes in currents in certain local areas. In the Caribbean, for example, there are upwellings in the region of the West Indies, near the Carioca Trench and Lake Maracaibo, and between Barranquilla and Jamaica. These cause a local enrichment of the water, considerable phytoplankton and zooplankton production, and a nursery ground for small zooplankton between the West Indies and the Aves Bank. It may be possible in some of these areas to produce additional upwelling by such means as a thermal fission plant deep in the ocean which would cause upward movement of the water by convection, or by air-lift pumps in limited areas. However, we do not yet know enough to predict whether such an upwelling would produce a profitable increase in a fishery. We have recently found that most fish larvae require microscopic animals of a limited and varying size range, even though the adult fish may be plant-eaters. Thus, a certain larva at birth will need food between 5 and 10 microns; as it grows, it will need food from 20 to 50 microns and so on; but it may change its diet from animals to plants at certain stages. If these conditions are not met, the larva will die. Deep waters contain higher nutrient levels than surface waters. However, unless they contain a complete nutrient medium they may be useless for providing all the necessary conditions for a large fish population. For example, if they do not contain certain growth-factors or vitamins, they may fail to produce a crop of phytoplankton, even though all else is there. Even if a plankton culture is given all that is theoretically required, it will not always grow well; possibly some trace-element may be wanting. The most practical suggestion so far made is the possibility of pumping nutrient-rich oceanic deep water into atoll lagoons.[5]

For mariculture the greatest potential lies in farming our estuaries and bays. It is possible to produce some 2,000 lbs of animal protein per acre in some estuarine environments, as has been done in Taiwan.[6] Maybe an even higher figure could be reached with greater understanding of the ecology of estuaries.

However, it is difficult for a biologist to persuade the average chemist of the difficulties of micro-algal and zooplankton culture, and estimates of production are made on the assumption that all consumable materials will actually be consumed to produce phytoplankton, and thence fish. A prediction of this sort claimed that an acre of seawater at St. Croix in the Caribbean could produce 33,000 lbs of protein. Hitherto, the best recorded marine fish farm production is about 2,600 lbs/acre/annum in brackish water ponds in the Ganges estuary; these benefit from the fertilization by the Calcutta sewage effluents.[7] On the full commercial scale, the best results are those of Taiwan, where 2,000 lbs/acre/annum and slightly more are regularly harvested. The fish farmers use a combination of organic fertilizer, supplementary fodder for the fish when the natural growth of algal pasture is cropped down by the grazing and growing fish, and by stocking with a succession of size-groups of fish. The organic fertilizer, such as rice bran, compost, and chicken manure, stimulates a rich growth of algae, chiefly blue-greens, on which flourishes the *Chanos* or milkfish, the popular and well-priced product of the ponds. No terrestrial farmer could claim a like animal protein production of 1,500 to 2,000 lbs/acre/annum. Such highly-productive estuarine areas await exploitation in many parts of the world, and offer great possibilities.

Much higher fish crops have been reported from estuarine areas; but these are obtained by factory-farming methods, in which the fertility of the estuary plays no part. The caged fish are fed massive rations of scrap fish, prawns, and meat, or pelleted foods; and naturally can give very high crops when measured in terms of production per unit area.[8]

There are shrimp fisheries even in polluted estuaries such as at Sydney. The shrimps live on microscopic animals in the sediments, which in turn live on the microbes inhabiting the bottom. The phytoplankton is not important in estuaries and very few animals seem to eat it, preferring the microbes of the bottom, especially those which grow attached to all sorts of surfaces such as the sea-grasses. In the average estuary, especially in warmer waters, there is an abundant growth of such grasses (turtle-grass, eel-grass, swan-grass, etc.), all of which are related to such plants as the Iris, but have returned to a life in the sea. These form a very good substrate for the settlement of many microscopic algae, and may have an aggregate surface area larger than that of the sea-bottom on which they grow. When these grasses break up and die, they produce a great mass of humus and detritus, which is in turn an excellent medium for the growth of the bottom-dwelling microbes. The Japanese stimulate fish production by strewing stones and rocks on the sandy bottoms of such bays as Mururan and Hakodate. These afford attachment for seaweeds, which they claim may increase by tenfold the production of fish, since the weeds produce humus and also afford shelter to the young fish.

There always seems to be surplus material, especially of plant material, in estuaries, and it might be thought that the zooplankton and fish populations would increase to the point where they were consuming all the available material; but this does not happen. There must be some factor other than primary productivity that limits the fertility of an estuary. If this idea is accepted, then fertilization will not necessarily give us an increased crop. We have first to seek the cause of the failure of the system to provide maximum efficiency. An attempt to fertilize an oyster-growing area in Botany Bay was abortive, due to the unpredictability of water-movements in the bay, and the channeling of the fertilizer into unprofitable areas by physical and chemical conditions; much was also lost by uptake into unprofitable organisms. This was as much a typical sea-grass area as possible. Yet similar fertilization of the Norwegian oyster polls has long been practiced commercially, and more recently in Yugoslavia.

Another factor to be reckoned with in mariculture is the progressive increase in production in some artificial fishponds. Interference with the topsoil by erosion or bulldozing will ruin a field for many years, and then the farmer has to wait until the humus is restored and the necessary chemical and physical desiderata are attained. He will try to hasten this process by planting legumes to restore the nitrogen and humus, and by careful ploughing and harrowing to restore the tilth. A newly-made fishpond may be comparatively unproductive until deoxygenation due to decaying vegetable-matter has taken its course; but thereafter a rich humus builds up in the pond, assisted by the application by the farmer of organic fertilizers, and the fertility of the pond increases. It has been proved that the higher the humus content of the pond soil, the higher the fish crop.[9]

Nitrogen comprises most of the air, and atmospheric nitrogen is used on a vast scale as raw material for the industrial manufacture of nitrogenous fertilizers for use in agriculture; nitrogen is also fixed from the atmosphere by many plants such as blue-green algae and some bacteria. Phosphate does not occur in the atmosphere and it is of limited occurrence on the earth. We know that there is a vast waste of nutrient material in the sewage discharged into the sea, though much is salvaged by the marine phytoplankton. London's sewage can be traced far out into the North Sea as a tongue of rich phytoplankton growth; it is indirectly responsible for the high fertility of the southern North Sea, as compared with neighbouring areas of sea, and for many thousands of tons of fine sea fish.[10] New York's sewage has a similar effect.

To get even moderate production from estuaries, however, they must be prepared and kept for fish culture, and not shared with industry and shipping. This would mean the abandonment of the practice of using every estuary as a dumping-ground for toxic industrial wastes, and the abolition of docks and marinas in all but a few carefully selected and suitable areas. The cynical attitude of certain industrial groups towards, for example, the thermal pollution of South Biscayne Bay in Florida, shows that the future of the population will for many years be at the mercy of vested interests and industrial concerns. The discharge of crude sewage from the city of Los Angeles into the Pacific Ocean was another case of cynical disregard of the people's rights to a pure ocean; a proper tertiary treatment plant would have provided clean beaches, and a great deal of water for the Los Angeles Valley to raise the water table and lessen the need for water from Lake Mead. In this case, political interests pointed out that Arizona and Nevada might object to supplying so much water for Los Angeles, and the matter became a political issue. Another tragic instance is the overpowering lobby of oil interests opposing some of the efforts to obviate future oil spills on the Santa Barbara, Texas and Louisiana beaches. Such a spill in the Laguna Madre or Aransas Bay of south Texas could ruin the productivity of fish or shrimp farms there for ten years or more.

Therefore, while the production of food from the estuaries to feed the growing populations is a number one priority for us today, it is doubtful whether anything of real import will happen until our values are altered. However, conservation has become a popular cause, leading to over-compensation for previous neglect, which may mean the dethroning of some interests that are acting against humanity. The stress on communities, both human and marine, will force a change; and mariculture will play an important part in the economy of the future, even in the United States. In some other countries, estuarine fish farming has long been traditional in the Far East, and in parts of southern Europe mariculture is being practised, sometimes on a very large scale, a world total of over one million acres producing about 250,000 tons of fish and prawns.

In the estuaries we have an active biological system with a tremendous rate of turnover, one which does not seem to reach its full potential. Thus, even without too much effort, it should be possible to improve production in many of our estuaries, provided we get to work before they become irreversibly polluted. There, the bacteria will play a very important part. Many industrial wastes could be made biologically useful by fairly simple microbiological treatment. Fungi and bacteria are able to degrade the most refractory of biological products, albeit some of them rather slowly. Chitin, which requires boiling acids for its chemical decomposition, is degraded by bacteria and other microbes at such a rate that it does not accumulate on the sea bottom, although vast quantities are produced by crustaceans, probably several times their own weight a year. Hydrocarbons, both straight chain and ring types, that occur in natural and artificial petroleum products, are known to be attacked and degraded by certain bacteria, many of which occur in the seas. Even sewage bacteria are attacked and eaten by many protozoa, particularly the ciliates, and thus taken out of harm's way provided that they are available to the predators; in the case of sewage, it must be freed from the fats and oils which protect it physically against the approach of protozoa. One has, then, to provide the right kind of organism, bacteria or protozoa, in the right place at the right time, and in sufficient numbers to do the work. It is really an extension of the principle of the septic tank. A new study for the marine microbiologist lies here, a study which would explain many of the phenomena of marine chemistry, and form a first line of defence against pollution.

## NOTES


1. Harvey, H. W. 1934. 'Measurement of Phytoplankton Populations.' *J. Mar. Biol. Assn.* U.K. *19* (2): 761-3.
2. Marshall, S. M., and Orr A.P. 1928. 'The Photosynthesis of Diatom cultures in the Sea.' *J. Mar. Biol. Assn. U.K. 15*: 761-73.
3. Steeman-Nielsen, E. 1953. 'Measuring the production of the sea.' *The 'Galathaea' Deep Sea Expedition.* George Allen & Unwin, London: 53-64.
4. Jackson P. 1954. 'Engineering and Economic Aspects of Marine plankton Harvesting.' *J. du Conseil,* Exp. Mer. Copenhagen *20*: 167-74.
5. Isaacs J. D., and Schmitt W. R. J. du Conseil Exp. Mer Copenhagen *33*(1): 20-26.
6. Lin S. Y. 1968 *Milkfish Farming in Taiwan.* Rept 3. Taiwan Fisheries Research Institute.
7. Pakrasi B., Das P., and Thakurta S. 1964. *Culture of brackish-water fishes in impoundments in West Bengal.* Indo-Pacific Fish. Coun. 11. Tech Document 19.
8. Hickling C.F. 1971. *Fish Culture* 2nd edn. Faber & Faber, London.
9. Tang Y.A., and Chen T.P. 1966. 'A survey of the algal pastures of Milkfish ponds in Taiwan'. F.A.O. symposium on warm-water fish culture. Rome 1966.
10. Kalle K. 1953. 'Der Einfluss des englischen Küstenwassers duf den Chemismus der Wasserkorper in der südlichen Nordsee.' *Ber. Deutsch-Wiss. Komm. f. Meeresf.* XIII(2): 130-135.

# INDEX